*UNDERSTANDING AND LEARNING STATISTICS BY COMPUTER*

Doing experiments helps us to remember facts and appreciate theories. It is an indispensable part in many branches of science and can also be used to understand statistics. This book teaches you how to use small computers to perform statistical experiments. On one hand, it covers all the basic concepts of statistics, and on the other hand, it tells the reader how to evaluate his own newly developed statistical procedures that cannot be found from any existing books.

*About the authors*

**Mark C. K. Yang** received his B. S. in Electrical Engineering from National Taiwan University and his M. S. (Mathematics), Ph. D. (Statistics) from the University of Wisconsin, Madison. He is currently a Professor at the Department of Statistics, University of Florida. In the past, Professor Yang worked with the Redstone Arsenal, NASA, Department of Energy, Naval Research Laboratory, and the Bell Laboratories. He has published more than 30 articles in statistics and engineering journals.

**David H. Robinson** received his B. S. in Mathematics from Henderson State University and his M. S., Ph. D. (Statistics) from the University of Iowa, in Iowa City, Iowa. He is currently an Associate Professor in the Department of Mathematics/Computer Science at St. Cloud State University in St. Cloud, Minnesota. In the past, he has taught at the University of Florida in Gainesville, Florida and at Henderson State University in Arkadelphia, Arkansas. His publications include articles for the *Journal of the American Statistical Association* (JASA) and for *Psychometrika.*

**World Scientific Series in Computer Science**

Volume 1: Computer-aided Specification Techniques
by J Demetrovics, E Knuth & P Radó

Volume 2: Proceedings of the 2nd RIKEN International Symposium on Symbolic and Algebraic Computation by Computers
edited by N Inada & T Soma

Volume 3: Computational Studies of the Most Frequent Chinese Words and Sounds
by Ching Y Suen

Volume 4: Understanding and Learning Statistics by Computer
by M C K Yang & D H Robinson

Volume 5: Information Control Problems in Manufacturing Automation
by L Nemes

Volume 6: D C Flux Parametron
A new approach to Josephson junction logic
by E Goto & K F Loe

Series in Computer Science — Vol. 4

# UNDERSTANDING AND LEARNING STATISTICS BY COMPUTER

Mark C. K. Yang
*Department of Satistics*
*University of Florida*

David H. Robinson
*Department of Mathematics and Computer Science*
*St. Cloud State University*

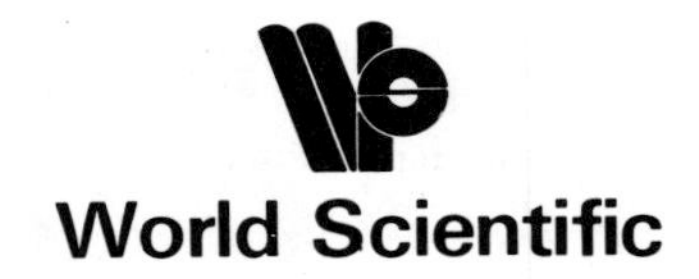

**World Scientific**

*Published by*

**World Scientific Publishing Co Pte Ltd.**
**P. O. Box 128, Farrer Road, Singapore 9128**
242, Cherry Street, Philadelphia PA 19106-1906, USA



**Library of Congress Cataloging-in-Publication Data**

Yang, Mark C. K.
Understanding and learning statistics by computer.

1. Statistics — Data processing. I. Robinson, David H. II. Title.
QA276.4.Y36 1986 519.5'028'5 85-29610
ISBN 9971-50-019-1
ISBN 9971-50-091-4 pbk



Printed in Singapore by Fu Loong Lithographer Pte Ltd.

# PREFACE

This book seeks to provide an introduction to statistics for those who are used to working with a computer or students who major in computer science. Though some of the commonly covered topics in most elementary statistics textbooks, such as the $t$, chi-square, and $F$ tests, have to be introduced, our emphasis is on how to use the computer, not only to facilitate statistical computation, but to understand statistics. With a computer at hand, one can easily discuss the statistical concepts that are seldom touched at the elementary level such as power, robustness, and efficiency of statistical procedures, and computer intensive techniques. We believe those concepts are no less important than the ones discussed in most elementary statistics books.

Since the emphasis is on the basic concepts, the mathematics is kept as simple as possible. However, some background in calculus is necessary in reading this book and we feel that this is not an unreasonable prerequisite for computer science majors. Again, only simple differentiation and integration are used.

There are programming exercises in every chapter of this book. They are simple in terms of programming complexity, but in order to do them one needs to know how to write arithmetic operations and connect subroutines. Any present-day microcomputer is adequate to do all the exercises in this book.

This book is short, again because the emphasis is on the basic concepts. Once these concepts are well understood, one can easily examine more complicated statistical procedures or develop their own when special applications arise.

The authors are indebted to the Department of Statistics, University of Florida and to Richard Scheaffer for the opportunity and encouragement during our teaching undergraduate statistics for computer science majors.

# PREFACE

# CONTENTS

Chapter 3. THE BINOMIAL DISTRIBUTION AND ITS APPLICATIONS TO SIMULATION



Chapter 4. COMPARISON OF STATISTICAL PROCEDURES



Chapter 5. ELEMENTARY STATISTICAL PROCEDURES

Chapter 1

# SAMPLING BY COMPUTER SIMULATION

## 1.1. A Problem Without a Definite Answer

One of the important purposes of gaining knowledge is to solve problems. Even with the best knowledge we have, the extent of the solution to a particular problem varies, from having a complete solution to being practically unsolvable. Statistical methods are tools to solve problems when a complete analytical solution is not possible, but a partial solution can be found.

For example, suppose we wish to find out the percentage of students who prefer a quarter system to a semester system in a certain university. One solution to this problem is to ask all the students in the university for their preference. If this is possible, the percentage is then known and this problem is completely solved. But suppose this university has 30,000 students and we only have the manpower to question at most 200 students. Then, strictly speaking, our information on the 200 students is too little to completely solve our problem. Suppose 150 out of the 200 students are in favor of a quarter system; then all we know for sure is that at least 150 out of the 30,000 students are in favor of a quarter system and at least 50 out of the 30,000 students are not in favor of the quarter system. Thus, the true proportion of students preferring a quarter system can be any number in the range from 150/30,000 to 29,950/30,000; that is, from 0.5% to 99.8%. This range, of course, is practically useless. But intuitively we know the percentage of students favoring a quarter system should be somewhat close to the percentage we observed, i.e.,

$$\frac{150}{200} = 0.75 = 75\% \quad .$$

The problem is to assess the accuracy of this 75%. How close is this 0.75 to the true proportion, say, $\theta$? If a "definite" answer about $\theta$ is required, then the solution could be very difficult, because the true $\theta$ is unknown and we cannot assess the closeness of 0.75 to an unknown quantity. Moreover, we know the closeness of our estimate to the true $\theta$ varies from one sample to the next. If we take another sample of 200 students, we will likely get a different estimate, which will be either closer or further from the true answer ($\theta$). But we may be able to make a statement such as

*The estimated proportion 0.75 is probably pretty close to the true proportion $\theta$.* (1.1.1)

The above statement has little scientific value because it does not specify how *"probably"* is 0.75 close to $\theta$, nor how *"close"* is 0.75 to $\theta$. Apparently, if one asked only 4 students and found three favorable to the quarter system, he might also make a statement such as

*The true proportion is probably pretty close to my estimate $3/4 = 0.75$.*

(1.1.2)

The meanings of (1.1.1) and (1.1.2) should be, of course, quite different due to the difference in the two sample sizes.

One important task in learning statistics is to quantify the statements (1.1.1) and (1.1.2). Eventually, we will be able to claim (1.1.1) precisely as the following:

*With at least 95% probability the difference between my estimate and the true proportion $\theta$ is smaller than 0.07.* (1.1.3)

or in the mathematical form

$$\Pr\{|\hat{\theta}-\theta| \leqslant 0.07\} \geqslant 0.95 \quad , \tag{1.1.4}$$

where $\Pr\{\ldots\}$ denotes a probability, and $\hat{\theta}$ denotes the estimated value of $\theta$. Before jumping from (1.1.1) to (1.1.3) or (1.1.4), we will try to quantify (1.1.1) step by step.

## 1.2. Population, Sample, and Estimates

From the previous section, we know our present goal is to find the accuracy of our estimate $\hat{\theta}$ from a sample. To formally define a "sample," we need to first define a population. A population is the target set for which we would like to know some characteristics. Thus, the population of the example in the previous section is all the students in that university. A sample is defined to be a *random subset* of the population, which means that each member of the population must have an equal chance to be included in the sample. Hence, the 200 students we questioned in the previous example is a (random) sample if every student has had an equal chance of being selected. The idea of *random* subset is important. For example, if we had asked 200 engineering students, we would not have a (random) sample from the whole student population, and our results would be biased toward the opinions of the engineering students.

One way to assess the accuracy of our estimate is to do an experiment. Since it takes quite an effort to randomly select 200 students and ask them questions, we will set up an experimental situation similar to that of taking a real sample in the student population. Let us buy 30,000 marbles and put them into an urn. The marbles are colored red and white to represent the students of two different opinions, and we will randomly take 200 marbles and record their colors. In this experiment, we actually know the true proportion of, say, red marbles. Let us assume that there are 21,426 red balls. Then the true proportion $\theta$ is 21,426/30,000 = 0.7142. Suppose we take 200 marbles and find that 145 of them are red; then our estimate is $\hat{\theta}$ = 0.725. In other words, the difference between the true and estimated proportions is known to be $|0.735 - 0.7142| = 0.0108$ for this sample. If we do not get tired of doing this experiment, we can collect samples of 200 marbles many, many times, each time replacing the 200 marbles back into the urn after we have drawn them. Suppose we perform this experiment 1,000 times and find the following results (Table 1.2.1).

Table 1.2.1

| Results | No. of times (out of 1,000) |
|---|---|
| $\lvert\hat{\theta} - \theta\rvert < 0.01$ | 225 |
| $0.01 \leqslant \lvert\hat{\theta} - \theta\rvert < 0.05$ | 646 |
| $0.05 \leqslant \lvert\hat{\theta} - \theta\rvert < 0.10$ | 124 |
| $0.10 \leqslant \lvert\hat{\theta} - \theta\rvert < 0.20$ | 4 |
| $0.20 \leqslant \lvert\hat{\theta} - \theta\rvert < 0.50$ | 1 |
| $0.50 \leqslant \lvert\hat{\theta} - \theta\rvert$ | 0 |

What does this result tell us? It tells us that roughly 22.5% of the time we should expect $|\hat{\theta} - \theta| < 0.01$, 64.6% of the time $|\hat{\theta} - \theta|$ should be between 0.01 and 0.05, etc. Moreover, we know that it is unlikely (only about 0.5% of the time) that we will observe a difference between $\hat{\theta}$ and $\theta$ of more than 0.1. Thus, based on our experimental results, we can say that, if we take a sample size of 200 out of a population of 30,000, and if the true proportion is around 70%, then our estimate has a very small chance, around 0.5%, of being in error by more than 0.1. Or we may write

$$\Pr\{|\hat{\theta} - \theta| \geqslant 0.1\} \cong 0.005 \quad . \tag{1.2.1}$$

To take 200 balls out of an urn one thousand different times is a tremendous amount of work. Maybe we can ask the computer to do this experiment for us. This will be the next section.

## 1.3. Taking a Sample by Computer Simulation

To represent a population with 30,000 elements in it is very easy for a computer. Thus, we may let the population described in the previous section be defined in an array $X$ as follows:

$$X(1) = X(2) = \ldots = X(21,426) = \text{RED}$$

$$X(21,427) = \ldots. = X(30{,}000) = \text{WHITE} \quad .$$

To make the notation simpler, we may use 1 to represent RED and 0 to represent WHITE. Thus, our storage array will be

$$X(1) = X(2) = \ldots = X(21,426) = 1, \quad \text{and}$$

$$X(21,427) = \ldots = X(30{,}000) = 0 \quad . \tag{1.3.1}$$

To use numerical values to represent the characteristics of a population usually greatly simplifies our mathematical formulation. These numerical values are usually called *random variables* before they are to be collected, and are called *data* after they are collected. Capital letters $X$, $Y$, $Z$ are usually used for random variables and small letters $x$, $y$, $z$ are usually used for data.

Now go back to the original computer experiment. We need to have a procedure that will tell the computer to pick an $X(k)$ at random. More precisely, each of the $X(i)$, for $i = 1, 2, \ldots, 30{,}000$, should have an equal chance of being chosen when we make a selection. To do this, we need to find a procedure that will give us a "random integer" between 1 and 30,000. Because we want our procedure to apply to other populations as well (with sizes different from 30,000), we need a general procedure to generate a random integer from 1 to $N$. The usual way to do this is to first generate a *random real number* $U$ between 0 and 1, then take the integer part of $N*U + 1$ as the required random integer between 1 and $N$. This procedure is continued until the desired number of samples is picked.

A random number $U$ between 0 and 1 is a number which is picked under the condition that each number between 0 and 1 has an equal chance to be picked. We say that $U$ is uniformly distributed between 0 and 1, and we write $U \sim U(0, 1)$.

Most computer software packages use a subroutine for generating such (uniform) random numbers. For example, the IBM/370 system uses the following iterative procedure to generate a sequence of random numbers $U_1$, $U_2$, $U_3, \ldots$.

*Algorithm 1.3.1*

START WITH AN ODD INTEGER $I_0$.

DEFINE $I_n = aI_{n-1} + c \pmod{m}$

LET $U_n = I_n/m$

OUTPUT RANDOM NUMBER $U_n$

SAVE $I_n$ FOR GENERATING THE NEXT $I_{n+1}$

Here $a$, $c$, and $m$ are specified integers, and mod $m$ is the modulus operation that outputs the remainder of an integer when it is divided by $m$. For example:

$$125 \pmod{100} = 25$$

and $$185 \pmod{43} = 13 \quad .$$

The IBM/370 system uses the following values for $a$, $c$, and $m$:

$$a = 7^5 = 16{,}807;\ c = 0;\ m = 2^{31} - 1 \quad .$$

This algorithm for generating uniform random numbers is a very common one among different computer languages and software packages, and is called the *linear congruential generator* (LCG). (When $c = 0$, the algorithm is also called a pure *multiplicative generator.*) The LCG generates a sequence of integer values ($I_n$) from 0 to $m - 1$ by using the mod $m$ operation. Each integer is transformed into the interval $[0, 1)$ by division by $m$ ($m$ is called the modulus). This produces a sequence of "pseudo-random" numbers $U_n$. They cannot be considered truly random since the sequence of numbers is completely determined by knowledge of $I_0$, $a$, $c$, and $m$. It is usually hoped that these pseudo-random numbers will behave in computer simulations just as if we were using truly random numbers.

Different values for $a$, $c$, and $m$ are possible, but it has been found that certain values produce preferable sequences of random numbers. For instance, we would not want the sequence of numbers to begin repeating, or "cycling," too quickly. Care must be exercised in the choice of values of $a$, $c$, and $m$ for this reason. For a small computer without a random number generator, the following $a$, $c$, and $m$ are found to be satisfactory when the LCG Algorithm 1.3.1 is used:

$$a = 25173; \quad c = 13849; \quad m = 65536$$

*Example 1.3.1:* Generate 2 random numbers using a linear congruential generator with $a = 25173$, $c = 13849$, $m = 65536$ and a starting value, usually called *seed*, $I_0 = 11$.

Solution:
$$\begin{aligned}
I_1 &= aI_0 + c (\mathrm{mod}\ m) \\
&= 25173 \times 11 + 13849 (\mathrm{mod}\ 65536) \\
&= 290752 (\mathrm{mod}\ 65536) \\
&= 28608 \\
U_1 &= 28608/65536 = 0.436523 \\
I_2 &= aI_1 + c\ (\mathrm{mod}\ m) \\
&= 25173 \times 28608 + 13849 (\mathrm{mod}\ 65536) \\
&= 720163033 (\mathrm{mod}\ 65536) \\
&= 53465 \\
U_2 &= 53465/65536 = 0.815811\ .
\end{aligned}$$

With our procedure to randomly sample from a population in a computer, the results described in Table 1.1.1 can be obtained in a matter of seconds.

## 1.4. Interpretation and Scope of Simulation Experiments

As we see from the previous section, computer simulation is equivalent to doing actual experiments. Thus, the interpretation of simulation results can be equivalent to the interpretation of experimental results. It depends inevitably on some unwarranted generalization. For example, suppose we have flown a particular airplane in test flights with full loads, empty loads, in rain and in clear weather for over one thousand miles. Then can we guarantee a safe flight in its first commerical flight when it has a 75% load, on a cloudy day with 500 miles to go? The experimenter probably would say that he is pretty sure about a safe trip, because the test flights have covered enough conditions that the present trip should be considered safe even though its particular conditions have not been met before. Similarly, if a simulation study has covered a broad range of practical circumstances, we will assume that its results are applicable in practice.

Take the example of the previous section again. We have done only one experiment with a sample size of 200, with a true red ball proportion $\theta = 0.7142$ in a population of size 30,000. Thus, we may not be sure whether our statement on the accuracy of the estimated proportion is valid under other circumstances. If we, however, have done experiments for a wide range of true $\theta$ 's, say $\theta$ from 0 to 1 with increments of 0.1, and if similar results as Table 1.1.1 have been obtained, then we will say that the error of our estimate being more than 0.1 has a very small probability (close to 0.005), no matter what is the true value of $\theta$. This conclusion, of course, has great value in practice, i.e., if we are satisfied with an accuracy of $\pm 0.1$ for $\theta$, then a sample of size 200 is sufficient.

## 1.5. Estimation of the Mean

The proportion of a certain characteristic, discussed in the previous sections, is an important parameter in a population. Examples are: the unemployment rate, the popularity of certain policies, the cure rate of certain drugs for a given disease, and the accident rate of certain types of drivers. But in other situations, an investigator is often interested in a quantitative description of a population, such as, the average family income of a community, the average height of the second grade students in an elementary school, and the average lifetime of a vacuum tube. These all involve *values* associated with each member in a population, not just a clear-cut characteristic such as red and white.

Suppose we wish to estimate the average height of the university students by taking a sample of 200 out of the university student population 30,000. What is a reasonable estimate of the population mean (average), $\mu$, which here is defined to be the average height of the 30,000 students? Intuitively, it should be the *sample mean,* i.e., if the heights of the $n = 200$ sampled students are

$$Y_1, Y_2, \ldots, Y_n \quad ,$$

then a reasonable estimate for $\mu$ is $\hat{\mu}$, or $\overline{Y}$, defined by

$$\hat{\mu} = \overline{Y} = \frac{1}{n} \sum_{i=1}^{n} Y_i \quad .$$

How good is this estimate? More specifically, we want a probability statement about the error $|\overline{Y} - \mu|$. We can use simulation techniques again.

Suppose we know the heights of 30,000 students or a similar file from the past year or from other universities. Then we can create a file in the computer with the 30,000 heights. Since we know the true mean $\mu$ in this case, we can always compute the error produced by the sample mean $\overline{Y}$. Consequently a table similar to Table 1.1.1 can be produced and a statement about the accuracy of the estimates $\overline{Y}$ can be issued.

To produce a data file of 30,000 heights may not be an easy matter. The task can be simplified if we can represent the 30,000 heights by a *histogram* or a bar chart (see Fig. 1.5.1).

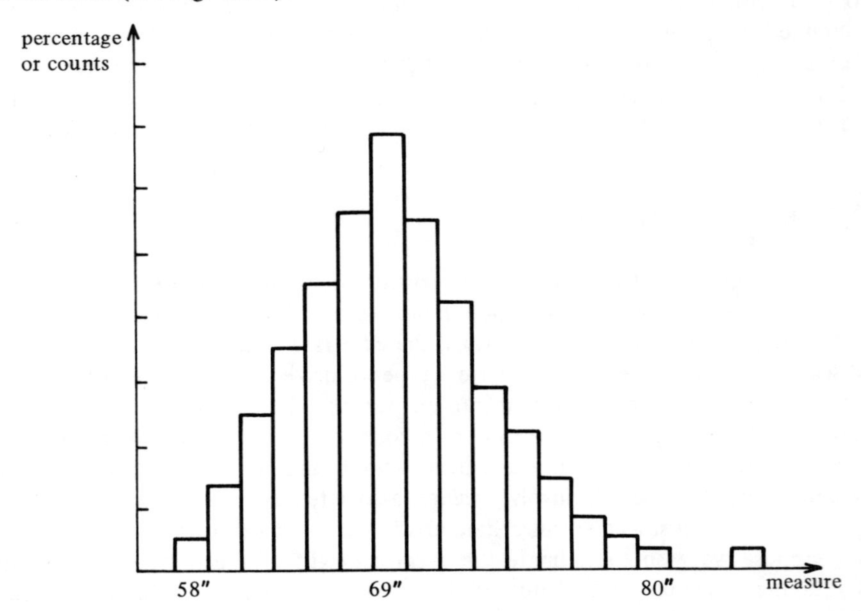


Fig. 1.5.1

Histograms are commonly used for describing the distribution of a population. They appear very often in newspapers and magazines. The diagram in Fig. 1.5.1 can be expressed symbolically by the following *frequency table*.

Table 1.5.1

| Class interval | Frequency (count) | Cum. Freq. | Relative freq. | Cum. rel. freq. |
|---|---|---|---|---|
| $(c_1, c_2]$ | $O_1$ | $N_1$ | $f_1$ | $F_1$ |
| $(c_2, c_3]$ | $O_2$ | $N_2$ | $f_2$ | $F_2$ |
| $(c_3, c_4]$ | $O_3$ | $N_3$ | $f_3$ | $F_3$ |
| . | . | . | . | . |
| . | . | . | . | . |
| . | . | . | . | . |
| $(c_{k-1}, c_k]$ | $O_{k-1}$ | $N_{k-1}$ | $f_{k-1}$ | $F_{k-1}$ |
| $(c_k, c_{k+1}]$ | $O_k$ | $N_k = 30{,}000$ | $f_k$ | $F_k = 1.00$ |
| TOTAL | 30,000 | | 1.00 | |

In Table 1.5.1,

$O_i$ = number of members in interval $(c_i, c_{i+1}]$

$N_i = O_1 + O_2 + \ldots + O_i$

= *cumulative count* of members from interval $(c_1, c_2]$ to $(c_i, c_{i+1}]$

= number of elements in $(c_1, c_{i+1}]$

$f_i$ = relative frequency in interval $(c_i, c_{i+1}] = O_i/N$

$$F_i = \text{cumulative relative frequency from } (c_1, c_2] \text{ to } (c_i, c_{i+1}]$$
$$= f_1 + f_2 + \ldots + f_i = N_i/N$$

$$N = \text{population size} = \sum_{i=1}^{k} O_i = N_k \quad . \tag{1.5.1}$$

To generate a data file for this population, we simply assign $X(1), X(2), \ldots,$ $X(N_1)$ into class 1, $X(N_1+1), X(N_1+2), \ldots, X(N_2)$ into class $2, \ldots,$ $X(N_{k-1}+1), \ldots, X(N_k)$ into class $k$. Their values will be uniformly assigned across the corresponding interval; i.e., if we let $\Delta_i$ = length of the $i$th interval $= c_{i+1} - c_i$, then

$$X(1) = c_1 + \frac{1}{O_1}\Delta_1, X(2) = c_1 + \frac{2}{O_1}\Delta_1, \ldots, X(N_1) = c_1 + \frac{O_1}{O_1}\Delta_1 = c_2$$

$$X(N_1+1) = c_2 + \frac{1}{O_2}\Delta_2, X(N_1+2) = c_2 + \frac{2}{O_2}\Delta_2, \ldots, X(N_2) = c_2 + \frac{O_2}{O_2}\Delta_2 = c_3$$

$$\vdots \qquad \vdots \qquad \vdots$$

$$X(N_{i-1}+1) = c_i + \frac{1}{O_i}\Delta_i, X(N_{i-1}+2) = c_i + \frac{2}{O_i}\Delta_i, \ldots, X(N_i) = c_{i+1}$$

$$\vdots \qquad \vdots \qquad \vdots$$

$$X(N_{k-1}+1) = c_k + \frac{1}{O_k}\Delta_k, X(N_{k-1}+2) = c_k + \frac{2}{O_k}\Delta_k, \ldots, X(N_k) = c_{k+1} \tag{1.5.2}$$

Once the data file for this population has been constructed, a sample can be drawn by using random numbers as described in Section 1.3, and the accuracy of the $\overline{X}$ with respect to the real population mean $\mu$ can be established with a table similar to Table 1.2.1.

## 1.6. Approximating the Distribution of a Large Data Set

In this section we will discuss an efficient method for storing the information about the distribution of a large set of data. Recall that in the last section we generated a data file representing the 30,000 heights of a population of university students. Since the only information used to generate the file was a histogram of the different heights, the actual values generated in the file were not the exact heights in the population, but merely conformed to the same histogram as the given population of 30,000 measurements. Such a file is useful for simulating samples from the population of heights, but it is cumbersome to store all 30,000 values. We would like a method by which we can recover any value in the data file, without storing them all in an array.

Let the population be defined by the frequency table in Table 1.5.1, and the population file created according to (1.5.2). Then the procedure of obtaining a piece of datum is to

*Algorithm 1.6.1*

(1) pick a random number $U$

(2) take the integer part of $N*U+1=J$

(3) pick the element $X(J)$ .

If we wish to determine which class interval $X(J)$ belongs to, we need to find location $J$ among the cumulative frequencies. According to (1.5.2), we know that $X(J)$ belongs to cell $i$, if and only if

$$N_{i-1} < J \leqslant N_i \quad . \tag{1.6.1}$$

Thus, for any population size $N$, one can show that (1.6.1) is equivalent to

$$\frac{N_{i-1}}{N} \leqslant U < \frac{N_i}{N} \quad ,$$

or

$$F_{i-1} \leqslant U < F_i \quad .$$

Hence, with the cumulative frequency table, we can locate the class interval to which our datum belongs, without referring back to the population. Once we have located the class interval for our observation, $X$, the actual value of $X$ should be

$$X = c_i + \frac{[N*(U-F_{i-1})+1]}{O_i} \Delta_i$$

$$\cong c_i + \frac{U-F_{i-1}}{f_i} \Delta_i \quad , \tag{1.6.2}$$

where $i$ denotes the index of the interval that $X$ belongs to, and $f_i$ is the relative frequency defined in Table 1.5.1. Thus, Algorithm 1.6.1 can be replaced by the following procedure without creating a population file.

*Algorithm 1.6.2*

(1) Pick a random number $U$

(2) Pick the class interval $i$, where $i$ satisfies $F_{i-1} \leqslant U < F_i$

(3) Let the datum be $X = c_i + \dfrac{(U - F_{i-1})}{f_i} \Delta_i$ .

Note that even the population size $N$ is not required in the sampling procedure defined by Algorithm 1.6.2. The population defined by its class intervals and their corresponding relative frequencies is an example of abstract discrete population. Such populations will be discussed in the next section.

Keeping track of all the relative frequencies $f_1, f_2, \ldots, f_k$, or equivalently, the cumulative frequencies $F_1, F_2, \ldots, F_k$ can also be difficult sometimes. We might rather replace the $F_i$'s by a smooth curve, representing an underlying abstract population of continuous measurement (see Fig. 1.6.1). Such populations will be discussed in the next chapter.

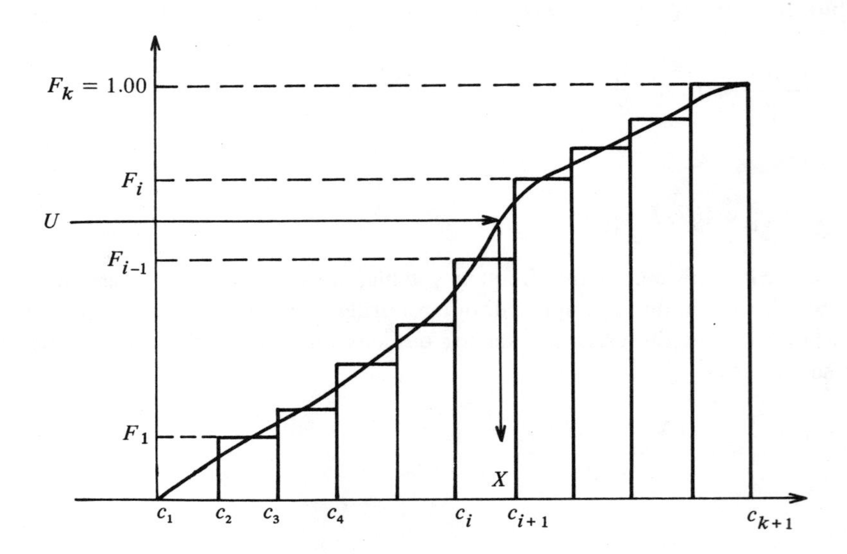


Fig. 1.6.1

### 1.7. Basic Elements in Probability

We have discussed probability without defining it. Actually probability is one of the basic terminologies or operations such as time, real numbers, and addition which cannot be explained by other simpler terms. We may say that probability is chance, but what is chance? In mathematics, we try not to define them but to state their properties. In probability we adapt the notations from the set theory. Let $S$, called the probability space or sample space, be the set that contains all the possible outcomes of an experiment (or a sample). All the subsets of $S$ are called events. For example, if we toss a fair die, then $S = \{1, 2, 3, 4, 5, 6\}$ contains all the possible outcomes of our experiment. The events such as

$$A \equiv \{1 \text{ will appear}\} = \{1\} \quad ,$$

$$B \equiv \{\text{It's an odd number}\} = \{1, 3, 5\}$$

$$C \equiv \{\text{It's 5 or 6}\} = \{5, 6\}$$

$$D \equiv \{\text{It's greater than 3}\} = \{4, 5, 6\} \tag{1.7.1}$$

are subsets of $S$. One basic property of probability is that if two events, A and B, are mutually exclusive, then the probability that one of them will happen is the sum of the two probabilities for A and B. In the above example, A and C are mutually exclusive, thus $\Pr\{A \text{ or } C\} = \Pr\{A \cup C\} = \Pr\{1, 5, 6\} = \Pr\{A\} + \Pr\{C\} = 1/6 + 1/3 = 1/2$. However, $\Pr\{C \text{ or } D\} \neq \Pr\{C\} + \Pr\{D\}$ because C and D have common elements.

It has been discovered by Kolmogorov in the 1930's that from only three axioms most desired properties of probabilities can be derived. They are:

(i) Let $S$ be the probability space, then $\Pr\{S\} = 1$.

(ii) For any event $A \subset S$, $0 \leqslant \Pr\{A\} \leqslant 1$ ,

(iii) For any two mutually exclusive events A and B (or $A \cap B = \phi$, the empty set),

$$\Pr\{A \cup B\} = \Pr\{A\} + \Pr\{B\} \quad ,$$

Moreover, for any sequence of mutually exclusive events $A_1, A_2, \ldots, A_n, \ldots$

$$\Pr\{\bigcup_{i=1}^{\infty} A_i\} = \sum_{i=1}^{\infty} \Pr\{A_i\} \quad .$$

From these axioms, one can easily derive the following theorems.

(a) $\Pr\{\overline{A}\} \equiv \Pr\{\text{not } A\} = 1 - \Pr\{A\}$ . (1.7.2)

(b) If $A \subset B$, then $\Pr\{A\} \leqslant \Pr\{B\}$ .

(c) For any two events A and B,

$\Pr\{A \cup B\} = \Pr\{A\} + \Pr\{B\} - \Pr\{A \cap B\}$, and

$\Pr\{A\} = \Pr\{A \cap B\} + \Pr\{A \cap \overline{B}\}$ . (1.7.3)

Another important concept in probability is that of dependent and independent events. Let $\Pr\{A|B\}$ denote the probability that A happens, given the condition that we know B has happened. Then it is quite intuitive that

$$\Pr\{A|B\} = \Pr\{A \cap B\}/\Pr\{B\} \quad . \tag{1.7.4}$$

Since B has happened, it is obvious that $\Pr\{B\} \neq 0$. In the die toss problem the probability of having an $A = \{1\}$ given it is odd (event $B = \{1, 3, 5\}$) is

$$\Pr\{A|B\} = (1/6)/(1/2) = 1/3 \quad .$$

This agrees very well with our intuition. We define that two events A and B are independent if $\Pr\{A|B\} = \Pr\{A\}$ ; i.e. the information on B does not change the probability of A. By (1.7.4) we see that A and B are independent if and only if

$$\Pr\{A \cap B\} = \Pr\{A\} \Pr\{B\} \quad . \tag{1.7.5}$$

Obviously the events defined in (1.7.1) are not independent of each other. But if two dies are to be tossed, then their outcomes are independent. The reader can easily verify (1.7.5) in this case.

One probability that is related to the sampling procedure in this chapter is the hypergeometric distribution. The typical problem of this type is that there are $N$ balls in an urn of which $R$ are red and $W = N - R$ are white. If $n$

balls are randomly drawn from the urn without replacement, then the probability of getting $r$ red balls is

$$\Pr\{X=r\} = \binom{R}{r}\binom{W}{n-r}\bigg/\binom{N}{n}, \tag{1.7.6}$$

where $X$ denotes the random variable that represents the number of red balls in the sample, and $\binom{a}{b}$ denotes the combinations of taking $b$ elements from $a$ distinctive elements. The derivation of (1.7.6) is based on the fact that there are $\binom{N}{n}$ ways to draw $n$ balls from $N$ balls, and there are $\binom{R}{r}\binom{W}{n-r}$ ways to draw $r$ red balls and $n-r$ white balls. Since every possible combination has the same chance of being picked, we have (1.7.6).

## Exercise 1

1.1 Write down an algorithm to generate die tosses. Suppose the first 3 random numbers you picked are 0.43341, 0.27638, and 0.74742, what are the first three tosses?

1.2 If $a = 2749$, $c = 447$, and $m = 19243$ are used for the linear congruential random generator, will they produce good random numbers?

1.3 How can one use a histogram to check whether a random number generator is uniform in (0, 1)?

1.4 A frequency table for the lengths of life of 417 incandescent lamps is shown below.

| Class intervals (h) | Frequency |
|---|---|
| 201 – 300 | 1 |
| 301 – 400 | 0 |
| 401 – 500 | 0 |
| 501 – 600 | 3 |
| 601 – 700 | 10 |
| 701 – 800 | 21 |
| 801 – 900 | 45 |
| 901 – 1000 | 91 |
| 1001 – 1100 | 85 |
| 1101 – 1200 | 80 |
| 1201 – 1300 | 44 |
| 1301 – 1400 | 23 |
| 1401 – 1500 | 9 |
| 1501 – 1600 | 3 |
| 1601 – 1700 | 2 |
| | 417 |

(i) Draw a histogram for this data.
(ii) Draw the cumulative relative frequency curve for this data.
(iii) If you use the standard method to draw a random sample of size 3 from this population, what are they, supposing the first 3 random numbers are 0.10480, 0.37570, 0.90229?

1.5 Compute the approximate mean of the above data.

1.6 The IBM scientific subroutine uses the following subroutine (in FORTRAN) to generate random numbers. What is the possible merit of using $m = 2^{31} = 2147483648$?

```
      SUBROUTINE RANDU (IX, IY, YFL)
      IY = IX * 65539
      IF(IY) 5, 5, 6
5     IY = IY + 2147483647 + 1
6     YFL = IY
      YFL = YFL * 0.4656613E - 9
      RETURN
```

1.7 (*programming exercise*) Let an urn contain $N$ marbles, of which $pN$ are white and the rest are red. We want to take a sample size $n$ and use the sample proportion

$$\hat{p} = (\text{number of white balls in the sample})/n$$

as the estimate of $p$. Use computer simulation to estimate $\Pr\{|\hat{p} - p| \geqslant 0.1\}$ by 1000 simulations for each of the following cases. (Let $p = 0.42$.)

| | $n$ | | | |
|---|---|---|---|---|
| | 20 | | 50 | |
| $N$ | w.r. | w/o.r. | w.r. | w/o.r. |
| 100 | ________ | ________ | ________ | ________ |
| 500 | ________ | ________ | ________ | ________ |
| 1000 | ________ | ________ | ________ | ________ |

In the above table "w.r." represents taking a sample with replacement, and "w/o.r." without replacement. Comment on the accuracy of $\hat{p}$ as effected by the following three factors: (i) the population size $N$, (ii) the sample size $n$, and (iii) the sampling procedures "w.r." and "w/o.r.".

1.8 Write down the sample space of the toss of two fair dice. Express the following events in set theory notation and find their probabilities.
(a) A = The sum of the two faces is 5.
(b) B = One of them is 1.
(c) C = The sum of the two faces is smaller than 5.
(d) D = 6 does not appear.

1.9 In Ex. 1.8 find (a) $\Pr\{A|B\}$ and (b) $\Pr\{C|D\}$ .

1.10 Show (1.7.3) and use it to show that

$$\Pr\{A\} = \Pr\{A \mid B\} \Pr\{B\} + \Pr\{A \mid \bar{B}\}\Pr\{\bar{B}\} \quad .$$

1.11 A data set contains 13,000 incomes.
(a) If I randomly pick one of them, what is the probability that the one I pick is the highest?
(b) If I randomly pick 100 of them, what is the probability that I pick the highest income?

1.12 A central processing unit (CPU) can simultaneously process three terminals. From past experience, the memory requests from the terminals during the day have the following distribution.

| | No Request | 15K (Word Processing) | 12K (BASIC) | 5K (Calculator) |
|---|---|---|---|---|
| Probability | 0.3 | 0.4 | 0.2 | 0.1 |

Suppose the CPU has memory size 32K and the terminal requests are independent.
(a) Find the probability that the requests will overload the CPU at a given time.
(b) If you wish to do word processing at one empty terminal, what is the probability that you cannot log on because of lack of memory space?

## Chapter 2

# COMMONLY USED POPULATIONS AND THEIR GENERATION

## 2.1. Abstract Discrete Populations

Even though all data sets which we encounter in practice are finite in size, we often consider them to be representative of a larger set of data, called the abstract population, which can be considered infinite in number. The 200 students questioned in Section 1.2. (or alternatively, the 200 marbles drawn from the urn) can be considered a representative sample of an *infinite* population, of which a fixed proportion $\theta$ is of one type, while the proportion $1-\theta$ is of the other type. The measurements of heights discussed in the last sections of Chapter 1 can similarly be considered as a sample of heights from an infinite population of possible heights. Such an assumption concerning the population is, of course, never true in practice, but it is helpful in setting up models for the *distribution,* or relative frequencies, of the different responses or measurements.

For example, when the real population of interest is the opinions of the 30,000 university students, if we instead consider the abstract population to be an infinite number of 0's and 1's (representing the two different possible opinions), then the only characteristic we need to know about the (abstract) population is the proportion of 0's and the proportion of 1's in the population. We represent the two proportions by $1-\theta$ and $\theta$, respectively. We can then list the relative frequencies and cumulative relative frequencies in a table such as Table 2.1.1. The relative frequencies of the two possible values in the population we call the probabilities $p_x = \Pr(X=x)$. That is, $p_0 = \Pr(X=0) = 1-\theta$ and $p_1 = \Pr(X=1) = \theta$. The cumulative relative frequency, or *cumulative distribution function* CDF, is found by accumulating (or summing) all the probabilities for points less than or equal to $x$. Thus

$$F(x) = \Pr(X \leqslant x) = \sum_{\{y: y \leqslant x\}} p_y \qquad (2.1.1)$$

Table 2.1.1

| Value ($X$) | Probability $p_x$ | CDF $F(x)$ |
|---|---|---|
| 0 | $1 - \theta$ | $1 - \theta$ |
| 1 | $\theta$ | 1 |

An abstract population in which there are only a finite, or at most countable, number of different possible values for $X$ is called a *discrete population.* The variable $X$ is then called a discrete random variable. The population of student opinions concerning the semester system is a discrete population since there are only two possible values for $X$, 0 and 1.

In general, for abstract discrete populations, we have a set of possible values for the random variable $X$ which we call the *support* $S$. For each value $x$ in $S$, there is an associated probability $p_x = \Pr(X = x)$. These probabilities must conform to certain laws of probability. In particular, we must have

(i) $0 < p_x \leqslant 1$ for each $x \in S$, and

(ii) $\displaystyle\sum_{\text{all } x \in S} p_x = 1$ . (2.1.2)

The CDF $F(x)$ is found by $F(x) = \displaystyle\sum_{\{y \in S: y \leqslant x\}} p_y$, for any value of $x$ along the real line. $F(x)$, by definition, gives the probability that the random variable $X$ takes on a value less than or equal to the number $x$. Note that $F(x)$ behaves like a step-function for discrete random variables. "Jumps" in the graph of $F(x)$ occur precisely at values $x$ in the support $S$. The height of the jump at a point $x \in S$ is equal to $p_x$. As an example, consider the random variable whose distribution is given in Table 2.1.2.

Table 2.1.2

| $X$ | $p_x$ | $F(x)$ |
|---|---|---|
| 0 | 1/4 | 1/4 |
| 1 | 1/2 | 3/4 |
| 2 | 1/4 | 1 |

The CDF is graphed in Fig. 2.1.1.

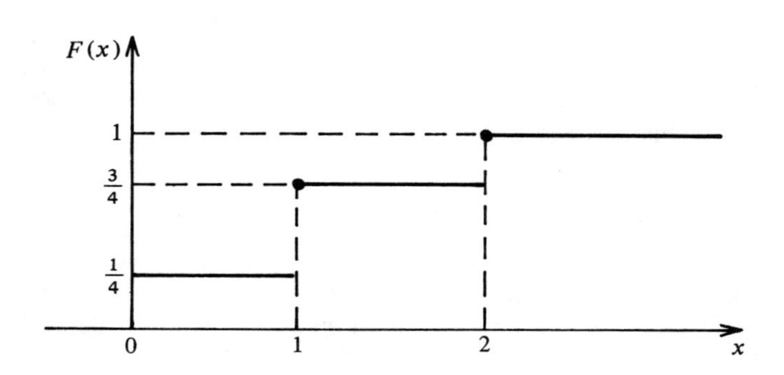


Fig. 2.1.1

## 2.2. Some Common Discrete Distributions

In this section we introduce some of the common probability distributions for discrete random variables, which we will use in later sections.

*1) Discrete uniform distribution*

Suppose that $S$ is some finite set of real numbers $x$. Then a random variable $X$ has a discrete uniform distribution over the support $S$ if the probabilities $p_x$ are all equal for each $x \in S$. That is, if $S$ contains $n$ different possible values for $X$, then $p_x = 1/n$ for each value. Two examples of discrete uniform random variables follow:

(a) Consider rolling a fair die and let $X$ denote the outcome (on the upper face) of a roll. Then we may consider an abstract population consisting of the support $S = \{1, 2, 3, 4, 5, 6\}$ with $p_x = 1/6$ for each $x$ in $S$. This is a discrete uniform distribution over the first six positive integers.

(b) Suppose a mechanism is able to "randomly" generate the digits 0 to 9. If the generation is done strictly at random, then each of the digits should be equally likely to occur at any point, and the probabilities $p_x$ assigned to each digit should be 1/10. Here $S = \{0, 1, 2, 3, 4, 5, 6, 7, 8, 9\}$ .

*2) Bernoulli distribution*

A Bernoulli distribution is one with only two possible values in the support: $S = \{0, 1\}$. We often think of such a random variable as resulting from an experiment with two possible outcomes: success ($X = 1$) or failure ($X = 0$). Usually $\Pr(X = 1) = p_1$ is denoted by $p$ (or $\theta$), with $\Pr(X = 0) = p_0$ denoted by $q = 1 - p$ (or $1 - \theta$). The abstract population of 0's and 1's discussed in Sec. 2.1 represents a Bernoulli distribution. (This distribution was named after the French mathematician Jacob Bernoulli.)

*3) Binomial distribution*

If we perform a sequence of $n$ Bernoulli experiments, each independent of the others and having the same probability of success $p$ for each experiment, then we might count $X$ = total number of successes in the $n$ Bernoulli trials. Such a random variable $X$ has a binomial distribution, denoted by $X \sim \text{Bi}(n, p)$. It can be shown (see Ex. 2.2), that the formula for $p_x$, for any $x \in S = \{0, 1, 2, \ldots, n\}$, is

$$p_x = \binom{n}{x} p^x (1 - p)^{n-x} \quad , \qquad x = 0, 1, 2, \ldots, n \quad , \qquad (2.2.1)$$

where $\binom{n}{x} = n!/(x!(n-x)!)$ is the binomial coefficient. For the example using the abstract population of student opinions about the semester system, if $X$ = the number of students, from a sample of 200, who favor a semester system, then we might consider $X$ to be a binomial random variable with $n = 200$ and unknown probability of success $p$. A table for $p_x$ for small $n$ is given in Appendix A.5.

*4) Geometric distribution*

Again suppose we perform a sequence of Bernoulli trials, but rather than fixing the number of experiments ($n$) to be performed, we choose to continue the trials only until the first success is observed. We might define the random variable

$X$ = number of trials needed to observe the first success.

Such a variable $X$ has a geometric distribution. It can be shown that the probabilities are given by

$$p_x = p(1-p)^{x-1} \quad , \qquad x = 1, 2, 3, \ldots \tag{2.2.2}$$

where $p$ is again the probability of success on any one trial. Note that the support $S = \{1, 2, 3, \ldots\}$ contains a countably infinite number of possible values for $X$.

*5) Poisson distribution*

Suppose we are measuring a counting-type of variable over a fixed period of time; e.g., we might count the number of cars arriving at a certain intersection during a day, the number of batch computer jobs submitted within a one-hour period, or the number of demands for service at a particular center during an 8-hour day. If we define a random variable $X$ to be the number of occurences in whatever random process we are observing, then often $X$ is modeled using the Poisson distribution (see Ex. 2.3). Such a distribution has probabilities given by

$$P_x = \frac{e^{-\lambda} \lambda^x}{x!} \quad , \qquad x = 0, 1, 2, \ldots \tag{2.2.3}$$

where $\lambda$ is the average rate of occurrence for the process for the period of time specified. For example, if batch jobs are submitted at the average rate of 10 per hour, then we can use $\lambda = 10$ in finding probabilities concerning

$X$ = number of batch jobs submitted within one hour.

Thus, for example,

$$\Pr(X = 5) = \frac{e^{-10}\, 10^5}{5!} = 0.038 \quad .$$

Note that any non-negative integer value for $X$ is theoretically possible, so again the support $S$ is countably infinite.

*Remark:* If $X$ is measured over a period of time $t$, and $\lambda$ is the average rate of occurrence per *unit* time, then we must use $\lambda^* = \lambda t$ in computing the probabilities for $X$. For instance, let $X$ = number of batch jobs submitted within a two-hour period. Then $\lambda^* = \lambda t = 20$; i.e., on the average, 20 batch jobs are submitted every 2 hours. Then

$$p_x = \frac{e^{-20}\,20^x}{x!} \qquad \text{for } x = 0, 1, 2, \ldots$$

## 2.3. Abstract Continuous Populations

Many times measurements are made where it is not conceivable to list all the possible values for the random variable in advance. Rather, it is often known that the variable could take on any value in an interval (possibly infinite in length) along the real line. Hence, the support $S$ consists of all values within this interval, of which there are uncountably many. For example, we might consider an abstract population based on the example of Section 1.5, where the random variable $X$ is the height of a university student. Considering the abstract population to be infinite in size, there are an uncountably infinite number of possible values for $X$. In fact, if $X$ is measured in inches, we might consider any value of $X$ between 50 and 84 inches to be in the support for the population (allowing for the possibility of some very short or very tall university students!). Such a population is said to have a *continuous* distribution, with the support contained in some continuous interval along the real line.

Since there are so many different possible values of $x$ for a continuous distribution, it is no longer possible to assign each separate value a positive probability in such a way the probabilities still add up to one. Hence we must settle for assigning probabilities to *intervals* within the support. Such probabilities are determined by a smooth curve, known as the *probability density function* (PDF).

The PDF of a continuous distribution has the following properties (we will use $f(x)$ to denote the PDF):

(i) $f(x) > 0 \qquad \text{for } x \in S\,,$

$\quad f(x) = 0 \qquad \text{for } x \notin S$

(ii) the area under $f(x)$ over the interval $(a, b)$ is equal to $\Pr(a < X < b)$; i.e.,

$$\int_a^b f(x)\,\mathrm{d}x = \Pr(a < X < b) \quad .$$

In most applications we are able to find probabilities such as this either by direct integration, or through the use of selected tables or computer approximations (numerical integration).

Certain consequences of these properties can be noted immediately. First we see that since the total probability for all values of $X$ must equal to one, then

$$\int_{\text{all } x \in S} f(x)\text{d}x = 1 \quad . \tag{2.3.1}$$

Thus the total area under the curve $f(x)$ must equal one. Secondly, the probability assigned to any one point $y \in S$ is given by

$$\int_y^y f(x)\text{d}x = 0 \quad .$$

Thus, $\Pr(X = x) = 0$ for each single point $x \in S$.

For example, consider the random variable $U$ which we generated using the LCG in Section 1.3. $U$ is said to have a continuous uniform distribution between 0 and 1, denoted by $U \sim U(0, 1)$, if the PDF for $U$ is constant over that interval. Thus we must have, because of (2.3.1),

$$f(x) = 1 \quad , \qquad 0 < x < 1$$

for the continuous uniform distribution on $S = (0, 1)$. This is the desired distribution for the random numbers we were generating. We noticed that we cannot *really* generate every possible value between 0 and 1. In particular, our LCG will only generate values of the form $k/m$, where $k$ is an integer from 0 to $m - 1$, and $m$ is the modulus of the LCG. However, these values are evenly spaced throughout the interval (0, 1) and it is hoped that they will suffice (for large $m$) as uniform random numbers. The limitations of our algorithm prevent us from generating *every* possible number.

The CDF of a continuous distribution is defined by $F(x) = \Pr(X \leqslant x)$, just as it was for discrete distributions. However, because of the difference in the way probabilities are calculated for discrete and continuous distributions, we have for continuous distributions

$$F(x) = \int_{\{y \in S : y \leqslant x\}} f(y)\text{d}y = \int_{-\infty}^{x} f(y)\,\text{d}y \quad .$$

Thus, the condition $y \in S$ is dropped in the second integral because we define $f(y) = 0$ for $y \notin S$. $F(x)$ is now represented graphically as the area under the PDF $f(y)$ to the left of the point $x$. Unlike the CDF for discrete distributions, the CDF for a continuous distribution will be a continuous function; there will be no "jumps." Note that to find $\Pr(a < X < b)$, it suffices to find $F(b) - F(a)$.

The CDF of the $U(0, 1)$ distribution is easily found to be

$$F(x) = \Pr(U \leqslant x) = \int_0^x 1\, du = x \quad , \qquad 0 < x < 1 \quad .$$

Thus we can graph $F(x)$ as in Fig. 2.3.1.

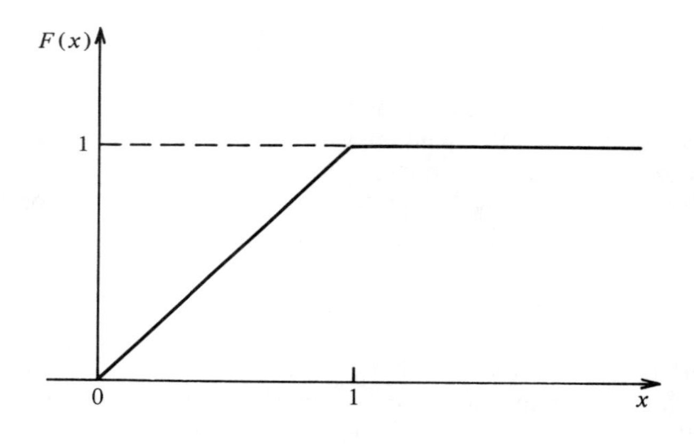


Fig. 2.3.1

Note that $F(x) = 0$ for all $x$ below the smallest value in the support, and $F(x) = 1$ for all $x$ larger than the largest value in the support.

The next section contains some of the common continuous distributions we will encounter later.

## 2.4. Some Common Continuous Distributions

### *1) Uniform distribution on (a, b)*

The random variable $X$ has a uniform distribution on the interval $(a, b)$, denoted $U(a, b)$, if its PDF is

$$f(x) = \frac{1}{b - a} \quad , \qquad a < x < b \quad . \tag{2.4.1}$$

Note that again $f(x)$ is constant for all $x$ in $S = (a, b)$, thus the name *uniform* distribution. Also note that the total area (or total probability) under $f(x)$ is one, which we knew must happen from (2.3.1). The CDF for the $U(a, b)$ distribution is easily found to be

$$F(x) = \frac{x-a}{b-a} \quad , \qquad a < x < b \quad .$$

*2) Standard normal distribution*

This is perhaps the most widely used probability distribution. Its main importance in statistics comes from its wide applicability and the Central Limit Theorem, which we will discuss in detail later. The random variable with a standard normal distribution is usually denoted by $Z$, the distribution itself is denoted by $N(0, 1)$ and together we use $Z \sim N(0, 1)$. Its PDF can be written as

$$f(z) = \frac{1}{\sqrt{2\pi}} e^{-z^2/2} \quad , \qquad -\infty < z < +\infty \quad . \tag{2.4.2}$$

Notice that the support for $Z$ in this case is the whole real line. This density function is the familiar "bell-shaped curve" referred to in many statistical applications. (See Fig. 2.4.1.)

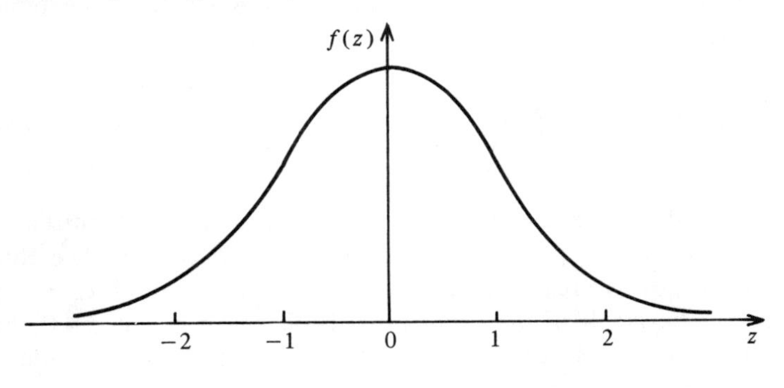


Fig. 2.4.1

Unfortunately there is no explicit expression for the CDF of the normal distribution:

$$\Phi(z) = \Pr(Z \leqslant z) = \int_{-\infty}^{z} \frac{1}{\sqrt{2\pi}} e^{-y^2/2} \, dy \quad . \tag{2.4.3}$$

(We usually use $\Phi(z)$ rather than $F(z)$ for the CDF of the $N(0, 1)$ distribution, but the meaning is the same.) Since this density function cannot be integrated directly, in order to find probabilities involving normal random variables, we must resort to tables giving us various percentile points of the distribution. For instance, from the $Z$-table given in Table A.1, we can see that $\Pr(-\infty < Z < 1) = .8413$ and $\Pr(-\infty < Z < 2) = .9772$. Since the PDF is symmetric and around 0, we also have $\Pr(-1 < Z < 0) = .3413$ and $\Pr(-2 < Z < 0) = .4772$. Certain percentiles of the $N(0, 1)$ distribution are used quite commonly in statistics. For example, the 0.95 percentile of the $N(0, 1)$ distribution is given by $z_{0.95} = 1.645$; that is, $\Pr(Z \leqslant 1.645) = 0.95$. Similarly, the 0.975 percentile is given by $z_{0.975} = 1.96$. Because of symmetry, we have the 0.05 percentile given by $z_{0.05} = -1.645$ and the 0.025 percentile given by $z_{0.025} = -1.96$. In general, we denote the $\alpha$ percentile of the $N(0, 1)$ distribution (in the left-hand tail) by $z_\alpha$ and the corresponding positive point in the right-hand tail by $z_{1-\alpha}$. That is, $\Phi(z_\alpha) = \alpha$ for any $0 < \alpha < 1$.

*3) Exponential distribution*

A common variable measured in engineering and related sciences is the waiting time until the first occurrence of a certain event. For example, we might define a variable $X =$ time until failure for a light bulb under test. Or we may have $X =$ waiting time until the next submission of a computer job on campus. Such random variables measuring time are often described with the exponential distribution, having PDF

$$f(x) = \lambda e^{-\lambda x} \quad , \qquad x > 0 \ , \tag{2.4.4}$$

where $\lambda$ is a positive constant representing the average rate of occurrence (per unit time) of the event, just as in the Poisson distribution. Notice that the support $S$ is all positive real numbers, since we could never observe negative waiting times. The graph of the exponential PDF is seen in Fig. 2.4.2. Notice that the PDF is highest at $x = 0$ and strictly decreasing from that point. The CDF for the exponential distribution is given by

$$F(x) = \int_0^x \lambda e^{-\lambda y}\, dy = 1 - e^{-\lambda x} , \quad x > 0 \ .$$

Graphically, $F(x)$ appears as in Fig. 2.4.3. The CDF approaches 1 asymptotically as $x$ gets larger, but never actually reaches 1.

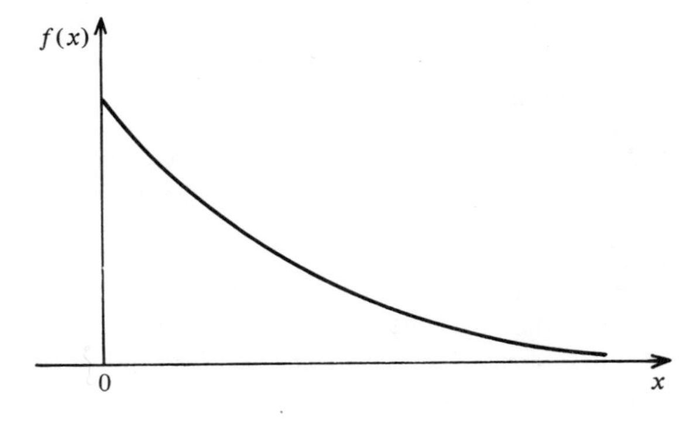


Fig. 2.4.2

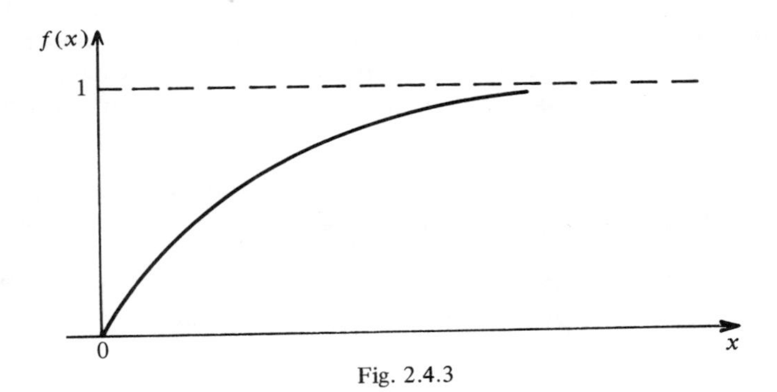


Fig. 2.4.3

As might be guessed, the exponential distribution is very much related to the Poisson distribution, which was discussed earlier. This relationship is also discussed in Ex. 2.13 and 14. Moreover, if we let $X(t) =$ number of jobs submitted in $[0, t]$, then $X(t)$ is called a *Poisson process.*

*4) Gamma (Erlang) distribution*

A random variable $X$ has a gamma distribution with parameters $\alpha$ and $\lambda$ if the PDF of $X$ is given by

$$f(x) = \frac{\lambda^{\alpha}}{(\alpha - 1)!} x^{\alpha - 1} e^{-\lambda x} \quad , \qquad x > 0 \quad .$$

Here $\alpha$ and $\lambda$ are positive constants which determine the shape of the distribution. In common applications, $\alpha$ will be integer-valued, in which case the distribution is also sometimes called an Erlang distribution, named after one of the early engineering pioneers for Bell Telephone. A gamma distribution (with integer-valued $\alpha$) commonly arises as follows:

> Let $X$ = waiting time until the $\alpha$th arrival or occurrence in some process (such as phone calls coming to a switchboard).

As with the exponential distribution, $\lambda$ represents the average rate of arrival per unit time. Notice that if $\alpha = 1$, the PDF of the gamma distribution reduces to that of the exponential distribution, as it should since $X$ then represents the time until the first arrival (or occurrence). Non-integer values of $\alpha > 0$ are also possible, though less common. In these cases we no longer have the physical interpretation of waiting time until the $\alpha$th arrival. The PDF in such cases uses the more general gamma function $\Gamma(\alpha)$ in place of $(\alpha - 1)!$.

The CDF of the gamma distribution with integer $\alpha$ can be found using integration by parts. Alternatively, since such integration is often quite involved, we can also use a relationship with Poisson probabilities. For example, to find the CDF $F(x)$ for a gamma distribution with $\alpha = 3$, we must find $\Pr(X \leqslant x)$, where

> $X$ = waiting time until the 3rd arrival (since $\alpha = 3$), with arrivals occurring at the average rate of $\lambda$ per unit time.

Then $X \leqslant x$ if and only if, at time $x$, we have observed *at least* 3 arrivals. We can find Pr(3 or more arrivals by time $x$) using Poisson probabilities:

$$p_3 + p_4 + p_5 + \ldots$$

$$= 1 - (p_0 + p_1 + p_2)$$

$$= 1 - \left(\frac{e^{-\lambda x}(\lambda x)^0}{0!} + \frac{e^{-\lambda x}(\lambda x)^1}{1!} + \frac{e^{-\lambda x}(\lambda x)^2}{2!}\right)$$

Here we are using $\lambda^* = \lambda x$ in the Poisson probabilities. Other probabilities from the gamma distribution can be worked out in a similar fashion in the exercises. This procedure avoids the use of integration to find such probabilities.

*5) Beta distribution*

The beta distribution is a common one in statistics for variables which are bounded between 0 and 1. The form of the PDF for the beta distribution is

$$f(x) = \frac{(\alpha+\beta-1)!}{(\alpha-1)!(\beta-1)!} x^{\alpha-1}(1-x)^{\beta-1} \quad , \quad 0 < x < 1 \quad .$$

$\alpha$ and $\beta$ are positive constants which determine the shape of the distribution. Usually $\alpha$ and $\beta$ are integer-valued, though not always. (When they are not integers, the more general function $\Gamma(\cdot)$ replaces the factorials in the PDF.) When $\alpha = 1$ and $\beta = 1$, we see that the PDF reduces to that of the continuous uniform distribution. For $\alpha > 1$ and $\beta > 1$, the distribution is mound-shaped. For $\alpha < 1$ and $\beta < 1$, the distribution is U-shaped. Thus we see that the beta distribution is quite flexible in terms of having many different shapes. The CDF for the beta distribution can be found by integration for small-integer values of $\alpha$ and $\beta$, but in general it is quite complicated. (See Fig. 2.4.4 for different shapes possible.)

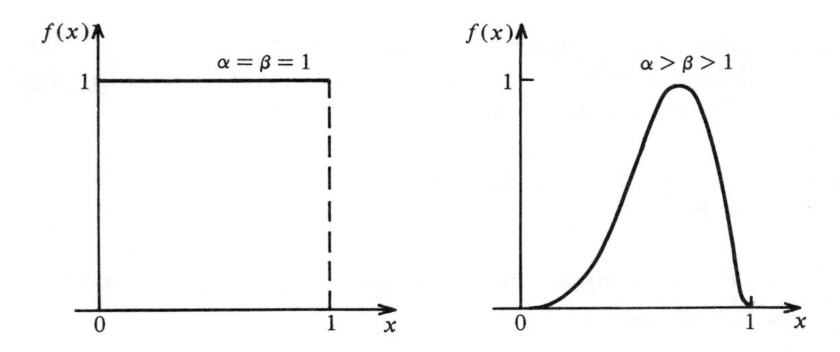


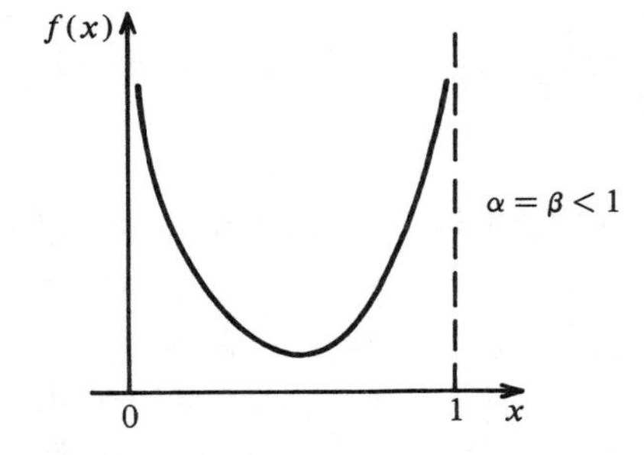


Fig. 2.4.4

## 2.5. Moments of Distributions

In general, we define the *kth moment* of a random variable $X$ to be the average of $X^k$ in the (abstract) population. We denote the $k$th moment of $X$ by $E(X^k)$. This average is calculated in one of two different ways, depending on whether the distribution of the population is discrete or continuous. For discrete distributions, we find the average by summing:

$$E(X^k) = \sum_{x \in S} x^k p_x \quad .$$

For continuous distributions, we replace the summation by an integral and $p_x$ by the PDF $f(x)$:

$$E(X^k) = \int_{x \in S} x^k f(x)\,\mathrm{d}x \quad .$$

For most practical purposes, $k$ will be some positive integer.

Of primary importance is the first moment of distribution ($k = 1$), called the *mean* of the distribution, and usually denoted by $\mu$. We have

$$E(X) = \sum_{x \in S} x p_x \qquad \text{for discrete distributions,}$$

$$= \int_{x \in X} x f(x)\mathrm{d}x \qquad \text{for continuous distributions.}$$

The mean of a distribution, or population, represents the average value of $X$ in the distribution, as we have seen previously in Section 1.5.

The second moment of a random variable $X$ is, by definition,

$$E(X^2) = \sum_{x \in S} x^2 p_x \qquad \text{for discrete distributions.}$$

$$= \int_{x \in S} x^2 f(x)\,\mathrm{d}x \qquad \text{for continuous distributions.}$$

The variance of a random variable $X$ is defined to be the average value of $(X - \mu)^2$ in the population, i.e., the average squared distance from $X$ to the mean $\mu$. The variance, denoted by $\sigma^2$, is thus a measure of dispersion in the population, and is found by

$$\text{Var}(X) = E[(X-\mu)^2] = \sum_{x \in S} (x-\mu)^2 \, p_x \quad \text{(discrete)} \quad ,$$

$$= \int_{x \in S} (x-\mu)^2 f(x) \, \mathrm{d}x \quad \text{(continuous)} \; .$$

It can also be shown that

$$\text{Var}(X) = E(X^2) - \mu^2$$

which is often easier computationally.

Another measure of dispersion in a population is the standard deviation $\sigma$, which is the square root of the variance.

To illustrate these calculations with an example, consider the discrete distribution presented in Section 2.1 which had probabilities

| $x$ | 0 | 1 | 2 |
|---|---|---|---|
| $p_x$ | 1/4 | 1/2 | 1/4 |

.

For this distribution, we find

$$\mu = E(X) = \sum_{x=0}^{2} x p_x = 1 \quad .$$

Also,

$$E(X^2) = \sum_{x=0}^{2} x^2 p_x = \frac{3}{2} \quad .$$

Thus, $\sigma^2 = E(X^2) - \mu^2 = 1/2$, and $\sigma = 1/\sqrt{2}$ .

As another example, one can easily find $\mu$ and $\sigma^2$ for the $U(0, 1)$ distribution.

$$\mu = \int_0^1 x \cdot 1 \, \mathrm{d}x = \left. \frac{x^2}{2} \right|_0^1 = \frac{1}{2}$$

$$E(X^2) = \int_0^1 x^2 \cdot 1 \, \mathrm{d}x = \left. \frac{x^3}{3} \right|_0^1 = \frac{1}{3}$$

$$\sigma^2 = E(X^2) - \mu^2 = \frac{1}{3} - \left(\frac{1}{2}\right)^2 = \frac{1}{12}$$

$$\sigma = 1/\sqrt{12} \quad .$$

Other examples can be worked out in the exercises.

The following are useful facts which can easily be proved for discrete or continuous random variables:

(i) If $E(X) = \mu$, then for any constants $a$ and b, $E(aX + b) = a\mu + b$.

(ii) If $\mathrm{Var}(X) = \sigma^2$, then for any constants $a$ and $b$, $\mathrm{Var}(aX + b) = a^2\sigma^2$.

The following can be proved using (i) and (ii) above.

(iii) If $X$ has mean 0 and variance 1, then for $Y = \sigma X + \mu$, $E(Y) = \mu$ and $\mathrm{Var}(Y) = \sigma^2$.

(iv) If $Y$ has mean $\mu$ and variance $\sigma^2$, then for $X = (y - \mu)/\sigma$, $E(X) = 0$ and $\mathrm{Var}(X) = 1$.
(Such a variable $X$ is said to be a *standardized* variable.)

It can be shown by technique of integration that the standard normal distribution $N(0, 1)$ has mean $\mu = 0$ and variance $\sigma^2 = 1$. If $Z$ has the $N(0, 1)$ distribution then it can be shown (using (iii)) that $Y = \sigma X + \mu$ has an $N(\mu, \sigma^2)$ distribution, i.e., a normal distribution with mean $\mu$ and variance $\sigma^2$. (See Fig. 2.5.1.)

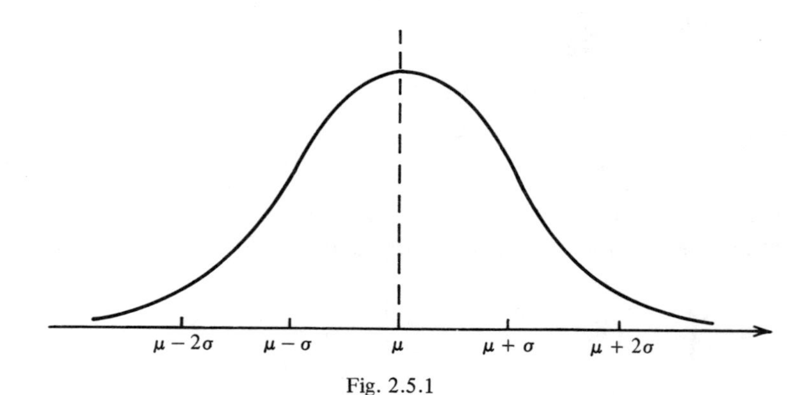


Fig. 2.5.1

Such a distribution has PDF given by the following formula:

$$f(x) = \frac{1}{\sqrt{2\pi\sigma}} e^{-(x-\mu)^2/2\sigma^2}, \qquad -\infty < x < +\infty .$$

Unfortunately, just as with the $N(0, 1)$ distribution, the $N(\mu, \sigma^2)$ distribution cannot be integrated directly to find probabilities or the CDF. Probabilities involving normal distributions are most easily handled by converting to the standard normal and using the appropriate $z$-tables. In other words if $X \sim N(\mu, \sigma^2)$, then $Z = (X-\mu)/\sigma \sim N(0, 1)$.

*Example 2.5.1:* Most biological variations are usually normally distributed. For example, the human height has almost a perfect normal distribution. Suppose an adult population has a normal distribution with mean 5′9″ and standard deviation 2.5″. Find the probability that a randomly chosen adult has height of more than 6 feet.

*Solution:* Let $X$ denote the height of a randomly chosen adult measured in inches. Then $X \sim N(69, 2.5^2)$. We wish to find $\Pr\{X > 72\}$. (Note 5′9″ = 69″ and 6′ = 72″). By standardizing $X$ we have

$$\begin{aligned}\Pr\{X > 72\} &= \Pr\left\{\frac{X-\mu}{\sigma} > \frac{72-\mu}{\sigma}\right\} \\ &= \Pr\left\{Z > \frac{72-69}{2.5}\right\} \\ &= \Pr\{Z > 1.2\} \\ &= 0.1151 \quad .\end{aligned}$$

Thus, the answer is 11.51%.

One important property of the normal population is that if $X_1, X_2, \ldots, X_n$ is a random sample from a normal population with mean $\mu$ and variance $\sigma^2$, then the sample mean $\overline{X}$ is still normally distributed. Its mean is still $\mu$, but its variance becomes $\sigma^2/n$, i.e.

$$\overline{X} \sim N(\mu, \sigma^2/n) \quad . \tag{2.5.1}$$

Another sample quantity that relates to the variance is called the sample variance which is defined by

$$\begin{aligned}S^2 &= \sum_{i=1}^{n} (X_i - \overline{X})^2/(n-1) \\ &= \sum_{i=1}^{n} X_i^2 - n\overline{X}^2)/(n-1)\end{aligned} \tag{2.5.2}$$

Obviously $S = \sqrt{S^2}$ (sometimes denoted by $s$) is called the sample standard deviation.

## 2.6. Approximating Some Useful Statistical Distributions

Several of the probability distributions which are particularly common and useful in statistical applications have the disappointing characteristic that their PDF's cannot be integrated in order to find probabilities. Prominent among these distributions are four of the distributions which are usually encountered in basic statistical applications: the normal distribution, (Student's) $t$ distribution, the chi-square distribution, and the $F$ distribution. Because probabilities are often needed from these distributions, statistical tables have been constructed for the purpose of making these probabilities available to the user of statistics. Different tables have been constructed in different ways, but one convenient method of presenting the tables is to give a list of $x$ values accompanied by a corresponding list of CDF values, $F(x) = \Pr(X \leqslant x)$. Then, for example, the 0.95 percentile of the distribution can easily be found by locating 0.95 in the list of values of $F(x)$ and reading the corresponding value for $x$.

In the following subsections we present numerical algorithms for the above four distributions which will, for a given value of $x$, compute $F(x) = \Pr(X \leqslant x)$. These algorithms will compute probabilities which are accurate to the third decimal place in most instances. The usefulness of tables which can be constructed from these algorithms will become more apparent in later chapters.

*1) Standard normal distribution*

We have previously seen that the PDF for the standard normal distribution is given by

$$f(z) = \frac{1}{\sqrt{2\pi}} e^{-z^2/2} \quad , \qquad -\infty < z < +\infty \quad .$$

An approximation for the CDF $\Phi(z) = \Pr(Z \leqslant z)$, for $z > 0$, is given by

$$\Phi(z) \cong 1 - \frac{1}{2(1 + a_1 z + a_2 z^2 + a_3 z^3 + a_4 z^4)^4} \quad , \qquad (2.6.1)$$

with $a_1 = 0.196854$, $a_2 = 0.115194$, $a_3 = 0.000344$, $a_4 = 0.019527$. If $z < 0$, then using the symmetry of the normal distribution, we have

$$\begin{aligned} \Phi(z) &= 1 - \Pr(Z > z) \\ &= 1 - \Pr(Z < -z) \\ &= 1 - \Phi(-z) \quad . \end{aligned}$$

The last expression involving $\Phi(\cdot)$ can be found using (2.6.1) since $-z > 0$ if $z < 0$.

*2) Chi-square distribution*

The PDF of a chi-square distribution with $\nu$ degrees of freedom is given by

$$f(x) = \frac{x^{\nu/2-1}\, e^{-x/2}}{2^{\nu/2}\,\Gamma(\nu/2)} \quad , \qquad 0 < x < +\infty \ .$$

$\Gamma(\cdot)$ is the gamma function, which is such that

$$\Gamma(a) = (a-1)!$$

for every positive integer $a$. Notice that the chi-square distribution is a special type of gamma PDF (discussed in Section 2.4), where the parameters are $\alpha = \nu/2$ and $\lambda = 1/2$. An approximation for the CDF $F(x)$ is given by

$$F_2(x) \cong \Phi(z)$$

where

$$z = \left[\left(\frac{x}{\nu}\right)^{1/3} + \frac{2}{9\nu} - 1\right]\sqrt{\frac{9\nu}{2}}$$

and $\Phi(z)$ is the CDF of the $N(0, 1)$ distribution.

*3) F distribution*

The PDF of the $F$ distribution with $\nu_1$ and $\nu_2$ degrees of freedom is given by

$$f(x) = \frac{\Gamma\left(\frac{\nu_1+\nu_2}{2}\right)\cdot\left(\frac{\nu_1}{\nu_2}\right)^{\nu_1/2}}{\Gamma\left(\frac{\nu_1}{2}\right)\cdot\Gamma\left(\frac{\nu_2}{2}\right)} \cdot \frac{x^{(\nu_1/2)-1}}{\left(1+\frac{\nu_1}{\nu_2}\,x\right)^{(\nu_1+\nu_2)/2}} \quad ,$$

$$0 < x < +\infty \ .$$

An approximation for the CDF $F(x)$ is

$$F_3(x) \cong \Phi(z)$$

where

$$z = \frac{\left(1 - \frac{2}{9\nu_2}\right)x^{1/3} - \left(1 - \frac{2}{9\nu_1}\right)}{\sqrt{\frac{2}{9\nu_2}x^{2/3} + \frac{2}{9\nu_1}}}$$

and $\Phi(z)$ is the CDF of the $N(0, 1)$ distribution.

*4) t distribution*

The PDF of the $t$ distribution with $\nu$ degrees of freedom is given by

$$f(x) = \frac{\Gamma\left(\frac{\nu+1}{2}\right)}{\sqrt{\pi\nu}\ \Gamma\left(\frac{\nu}{2}\right)} \cdot \frac{1}{\left(1 + \frac{x^2}{\nu}\right)^{(\nu+1)/2}} \quad , \qquad -\infty < x < +\infty \ .$$

Notice that the $t$ distribution, like the normal, is symmetric around $x = 0$. An expression for the CDF $F(x)$, for $x > 0$, is

$$F_4(x) = \frac{1}{2}[1 + F_3(x^2)] \quad , \qquad (2.6.2)$$

where $F_3(\cdot)$ is the CDF of the $F$ distribution with $\nu_1 = 1$ and $\nu_2 = \nu$ degrees of freedom. If $x < 0$, then using the symmetry of the $t$ distribution as we did with the normal, we have

$$\begin{aligned} F_4(x) &= 1 - \Pr(X > x) \\ &= 1 - \Pr(X \leqslant -x) \\ &= 1 - F_4(-x) \quad , \end{aligned}$$

which can be found by using (2.6.2), since $-x > 0$.

| *ID* | *F*1 | *F*2 | *X* | TABLE | COMPUTED | ERROR |
|---|---|---|---|---|---|---|
| 1 | 0 | 0 | 0.0000 | 0.5000 | 0.5000 | 0.0000 |
| 1 | 0 | 0 | 0.2500 | 0.5987 | 0.5987 | 0.0000 |
| 1 | 0 | 0 | 0.5000 | 0.6915 | 0.6917 | 0.0007 |
| 1 | 0 | 0 | 0.7500 | 0.7734 | 0.7734 | 0.0000 |
| 1 | 0 | 0 | 1.0000 | 0.8413 | 0.8411 | - 0.0002 |
| 1 | 0 | 0 | 1.2500 | 0.8944 | 0.8942 | - 0.0002 |
| 1 | 0 | 0 | 1.5000 | 0.9332 | 0.9333 | 0.0001 |
| 1 | 0 | 0 | 1.7500 | 0.9599 | 0.9602 | 0.0003 |
| 1 | 0 | 0 | 2.0000 | 0.9772 | 0.9774 | 0.0002 |
| 1 | 0 | 0 | 2.2500 | 0.9878 | 0.9878 | 0.0000 |
| 1 | 0 | 0 | 2.5000 | 0.9938 | 0.9937 | - 0.0001 |
| 1 | 0 | 0 | 2.7500 | 0.9970 | 0.9968 | - 0.0002 |
| 1 | 0 | 0 | 3.0000 | 0.9987 | 0.9984 | - 0.0003 |
| 1 | 0 | 0 | 3.2500 | 0.9994 | 0.9992 | - 0.0002 |
| 1 | 0 | 0 | - 1.5000 | 0.0668 | 0.0667 | - 0.0001 |
| 1 | 0 | 0 | - 2.5000 | 0.0062 | 0.0063 | 0.0001 |
| 1 | 0 | 0 | 3.5000 | 0.9998 | 0.9996 | - 0.0002 |
| 2 | 5 | 0 | 0.7270 | 0.7500 | 0.7480 | - 0.0020 |
| 2 | 5 | 0 | 2.0150 | 0.9500 | 0.9506 | 0.0006 |
| 2 | 10 | 0 | 0.2600 | 0.6000 | 0.6062 | 0.0062 |
| 2 | 10 | 0 | 0.7000 | 0.7500 | 0.7472 | - 0.0028 |
| 2 | 10 | 0 | 1.3720 | 0.9000 | 0.9009 | 0.0009 |
| 2 | 10 | 0 | 2.7640 | 0.990 | 0.9904 | 0.0004 |
| 2 | 15 | 0 | 0.2580 | 0.6000 | 0.6062 | 0.0062 |
| 2 | 15 | 0 | 0.6910 | 0.7500 | 0.7468 | - 0.0032 |
| 2 | 15 | 0 | 1.7530 | 0.9500 | 0.9515 | 0.0015 |
| 2 | 15 | 0 | 2.6020 | 0.9900 | 0.9905 | 0.0005 |
| 2 | 20 | 0 | 1.7250 | 0.9500 | 0.9517 | 0.0017 |
| 2 | 20 | 0 | 2.8450 | 0.9950 | 0.9951 | 0.0001 |
| 2 | 5 | 0 | - 0.7270 | 0.2500 | 0.2520 | 0.0020 |
| 2 | 10 | 0 | - 1.3700 | 0.1000 | 0.0994 | - 0.0006 |
| 2 | 15 | 0 | - 2.6000 | 0.0100 | 0.0096 | - 0.0004 |
| 3 | 10 | 0 | 2.1600 | 0.0050 | 0.0058 | 0.0008 |
| 3 | 10 | 0 | 3.2500 | 0.0250 | 0.0256 | 0.0006 |
| 3 | 10 | 0 | 6.7400 | 0.2500 | 0.2489 | - 0.0011 |
| 3 | 10 | 0 | 16.0000 | 0.9000 | 0.9008 | 0.0008 |
| 3 | 10 | 0 | 20.5000 | 0.9750 | 0.9754 | 0.0004 |
| 3 | 15 | 0 | 5.2300 | 0.0100 | 0.0104 | 0.0004 |
| 3 | 15 | 0 | 11.0000 | 0.2500 | 0.2465 | - 0.0035 |
| 3 | 15 | 0 | 25.0000 | 0.9500 | 0.9504 | 0.0004 |
| 3 | 20 | 0 | 8.2600 | 0.0100 | 0.0103 | 0.0003 |
| 3 | 20 | 0 | 12.4000 | 0.1000 | 0.0983 | - 0.0017 |
| 3 | 20 | 0 | 28.4000 | 0.9000 | 0.8999 | - 0.0001 |
| 3 | 20 | 0 | 34.2000 | 0.9750 | 0.9754 | 0.0004 |
| 4 | 3 | 5 | 3.6200 | 0.9000 | 0.8996 | - 0.0004 |
| 4 | 3 | 12 | 2.6100 | 0.9000 | 0.9009 | 0.0009 |
| 4 | 4 | 4 | 4.1100 | 0.9000 | 0.8991 | - 0.0009 |
| 4 | 10 | 12 | 2.1900 | 0.9000 | 0.9002 | 0.0002 |
| 4 | 1 | 3 | 10.1300 | 0.9500 | 0.9481 | - 0.0019 |
| 4 | 1 | 8 | 5.3200 | 0.9500 | 0.9517 | 0.0017 |
| 4 | 1 | 12 | 4.7500 | 0.9500 | 0.9522 | 0.0022 |
| 4 | 3 | 24 | 3.0100 | 0.9500 | 0.9508 | 0.0008 |
| 4 | 1 | 4 | 21.2000 | 0.9900 | 0.9887 | - 0.0013 |
| 4 | 1 | 8 | 11.2600 | 0.9900 | 0.9901 | 0.0001 |
| 4 | 3 | 9 | 6.9900 | 0.9900 | 0.9897 | - 0.0003 |
| 4 | 5 | 10 | 5.6400 | 0.9900 | 0.8897 | - 0.0003 |

Table 2.6.1. Comparison between tabled and computed $z$, $t$, chi-square, and $F$ probabilities (ID = 1 for $z$; 2 for $t$; 3 for chi-square; 4 for $F$; $F^1$, $F^2$, for degrees of freedom).

Table 2.6.1 presents a comparison between the approximated and accurate table values. The standard normal approximation has absolute error less than $2.5 \times 10^{-4}$. The other approximations tend to be very good if the degrees of freedom for chi-square and $t$ and the second degrees of freedom $\nu_2$ for $F$ are larger than 10. They are still reasonably good unless the degrees of freedom is extremely small such as 3 or less, which seldom produce any significant statistical conclusion anyway due to the scarcity of data. (See Chapter 5.)

## 2.7. Sampling from a Discrete Distribution

As we have already seen, most statistical simulations will require some random variables (actually, pseudo-random) to be drawn according to a specified probability distribution. They are usually called random variates (deviates) in contrast to random numbers which means $U(0, 1)$ random variables. If the distribution of interest is discrete, then the random variates we generate should only take on those values having positive probability of occurring. We would like for our random observations that we generate to assume these values with the same "long run" relative frequencies as specified by the probabilities. We have seen (Section 1.6) how to pick class intervals in a histogram according to the relative frequencies in the histogram. Generating numbers from a discrete distribution is done by essentially the same process, using continuous uniform random numbers in the interval (0, 1).

The procedure is best illustrated by example. Suppose we are given the probability distribution in the following table:

Table 2.7.1

| $X$ | 0 | 1 | 2 |
|---|---|---|---|
| $p_x$ | 1/4 | 1/2 | 1/4 |
| $F(x)$ | 1/4 | 3/4 | 1 |

We have been asked to generate random $X$-values from this distribution. Then we know we would like about 25% of the $X$'s we generate to equal 0, with about 50% equal to 1, and about 25% equal 2. (Of course, we could systematically generate $X$'s with the correct proportions such as

$$0, 1, 1, 2, 0, 1, 1, 2, 0, 1, 1, 2, \ldots,$$

but these would not appear random.) We can control these relative frequencies by using a sequence of $U(0, 1)$ random numbers. If we divide the interval $(0, 1)$ into three segments, $[0, .25)$, $[.25, .75)$, $[.75, 1)$, then for each uniform number we observe in the interval $[0, .25)$ we can assign the $X$-value of 0. This will result in approximately 25% of our $X$-values being 0. Similarly, we can assign the $X$-values of 1 for every uniform number falling in the interval $[.25, .75)$, and the $X$-value of 2 for every uniform number in the interval $[.75, 1)$.

Compare this procedure with the method for choosing class intervals in Section 1.6, using a histogram consisting of 3 class intervals, with $f_1 = .25$, $f_2 = .50$, $f_3 = .25$. Then $F_1 = .25$, $F_2 = .75$, and $F_3 = 1$, and we see that the procedures are actually equivalent. We can state our general rule in terms of the CDF $F(x)$ as follows:

*Algorithm 2.7.1*

(i) Generate a $U(0, 1)$ random number $U$.

(ii) Find the first value of $X$ for which $F(X) > U$; i.e., $X = \min\{x: F(x) > U\}$.

(iii) This value of $X$ is our "random" variate.

This method of generating random $X$-values is called the *inverse CDF method* for discrete distributions. Graphically, the procedure is illustrated in Fig. 2.7.1.

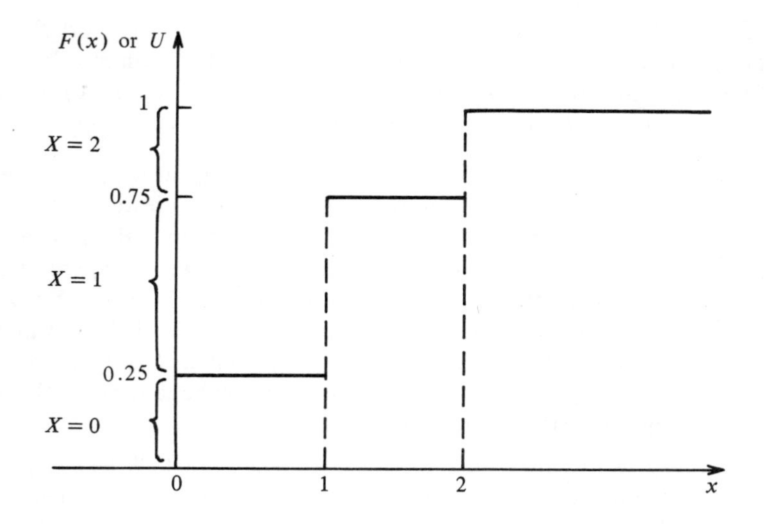


Fig. 2.7.1

Locate $U$ on the vertical axis, labelled $F(x)$. Move across horizontally until you meet a vertical line. This vertical line yields the proper value of $X$.

To carry out this procedure on a computer, we must store the values of $X$ and $F(x)$ in arrays. Let $X$ have a support $S$ with $K$ values, which are stored in $X(1)$, $X(2), \ldots, X(K)$. (In the example above, $K = 3$.) Store the CDF $F(x)$ for these $K$ values in $F(1)$, $F(2), \ldots, F(K)$. Notice that $F(K) = 1$. We must search the array $F$ for the first element larger than $U$, and then output the corresponding value from the array $X$. The following flow chart illustrates the procedure for obtaining an $X$-value from a given $U$.

For a given $U$.

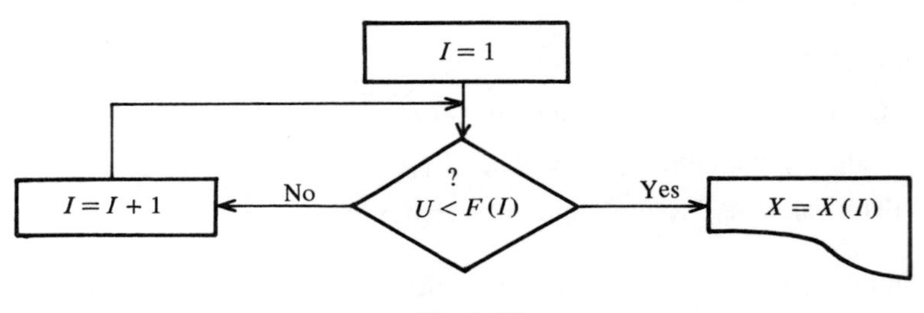


Fig. 2.7.2

A complication arises when $X$ is a random variable with support $S$ that is countably infinite, as with the Poisson distribution, for example. In such a case, it is clear that we cannot store *all* the values of $X$ and $F(x)$ in arrays. We must alter our procedure to one of the following methods.

1) Choose $K$ large enough so that $F(K)$ is almost equal to 1; for example, .99 or more. Then for any uniform random number greater than $F(K) =$ .99 (a rare occurrence), choose $X = X(K)$, rather than the proper value of $X$ which would fall outside the table. Thus we are approximating the desired distribution for $X$ with a distribution in which $X = X(K)$ is the largest possible value of $X$ which can be generated. The method is not totally correct because $X(K)$ is generated slightly more often than it should be, whereas no larger values can ever be generated. However, such an approximation should suffice in most cases.

2) Suppose $F(k)$ follows a mathematical formula so that all the $F(k)$, $X(k)$ can be computed by the computer. In this case one may first begin with a table $X(i)$ and $F(i)$ for $i = 1, 2, \ldots, K$ such that $F(K)$ is pretty close to 1. If a random number is generated and $U < F(K)$, then $X$ will be

chosen according to Fig. 2.7.2. Whenever $U \geqslant F(K)$, then $F(K+1)$, $F(K+2), \ldots$ will be generated sequentially until we reach the first $K'$ such that $U < F(K')$. We stop the process now and let $X = X(K')$. Since $F(i)$, $X(i)$ for $i > K$ are generated but not stored, the storage space of $F$ and $X$ will not produce any problem. However, because all the $F(i)$ and $X(i)$ have to be generated repeatedly for all $U \geqslant F(K)$ it can be very time consuming if $F(K)$ is not close to 1.

## 2.8. Sampling from a Continuous Distribution

When we wish to generate $X$-values from a continuous distribution with CDF $F(x)$, we apply the following procedure:

*Algorithm 2.8.1*

(i) Generate $U$ from the $U(0, 1)$ distribution.

(ii) Set $U = F(X)$, where $F(x)$ is the CDF for the desired distribution of the $X$'s.

(iii) Solve for $X$, obtaining $X = F^{-1}(U)$.

Values of the $X$ found in this way will have the correct distribution. This method of random variate generation is called the *inverse CDF method* for continuous distributions.

The reason for using this procedure is quite simple. Since $X = F^{-1}(U)$, we have

$$\begin{aligned} \Pr\{X \leqslant x\} &= \Pr\{F^{-1}(U) \leqslant x\} \\ &= \Pr\{U \leqslant F(x)\} = F(x) \quad . \end{aligned}$$

This $X$ satisfies the requirement that $\Pr\{X \leqslant x\} = F(x)$.

*Example 2.8.1:* Suppose that $X$ has an exponential distribution with PDF

$$f(x) = -e^{-\lambda x} \,, \quad x > 0 \ .$$

Then the CDF, as noted in Section 2.4, is

$$F(x) = 1 - e^{-\lambda x} \quad , \qquad x > 0 \ .$$

To generate random variates from this distribution, we first generate a sequence $U_1, U_2, U_3, \ldots$, from the $U(0, 1)$ distribution. For each $U_j$, we set $U_j = F(x) = 1 - e^{-\lambda x}$, and solve for $x$, obtaining

$$X_j = \frac{-\ln(1 - U_j)}{\lambda}$$

Those $X$-values calculated in this manner will follow the proper exponential distribution. For example, if we generated $U_1 = .3714$, $U_2 = .5893$, $U_3 = .8810$, and we desired 3 $X$-values from an exponential distribution with $\lambda = 2$, we would generate

$$X_1 = \frac{-\ln(1 - .3714)}{2} = .2321, \quad X_2 = .4449, \quad X_3 = 1.0643 .$$

(A slight simplification may be made at this point. If $U$ has a $U(0, 1)$ distribution, then $1 - U$ also has a $U(0, 1)$ distribution. Thus, everywhere $1 - U$ appears in the formula $X = F^{-1}(U)$, we may replace it by $U$, and the new $X$'s will still have the same distribution. In the above example, we may use $X_j = (-\ln(U_j))/\lambda$ to generate exponential random deviates. Given $\lambda = 2$ and the values for $U_1$, $U_2, U_3$ used above, we get

$$X_1 = \frac{-\ln(.3714)}{2} = .4952, \quad X_2 = .2644, \quad X_3 = .0633 .$$

This is a slightly faster calculation, making it preferable if we must perform it a large number of times.)

*Example 2.8.2:* Suppose that we would like to generate $X$'s having a distribution on $S = (1, +\infty)$ with PDF

$$f(x) = \frac{1}{x^2}, \qquad x > 1 .$$

Then the CDF is given by

$$F(x) = \int_0^x \frac{1}{y^2}\, dy = 1 - \frac{1}{x}, \qquad x > 1 .$$

(See Figs. 2.8.1 (a) and (b).) To generate $X$'s from this distribution, we set $U = 1 - 1/X$ and solve for $X: X = 1/(1-U)$. A sequence of $U$'s can thus be transformed into a sequence of random $X$'s from this distribution. (Again, according to the simplification suggested above, we may use $Z = 1/U$, which can be done slightly faster.)

A difficulty with this inverse CDF method is that we must have a closed form expression for the CDF $F(x)$, and then we must also be able to solve the equation $U = F(X)$ for $X$. These two requirements are not always satisfied. Some distributions, such as the $N(0, 1)$, have no exact expression for the CDF, other than as an integral. (Recall that the PDF of the $N(0, 1)$ distribution cannot be integrated analytically to find the CDF:

$$\Phi(z) = \int_{-\infty}^{z} \frac{1}{\sqrt{2\pi}} e^{-y^2/2}\, dy \quad .)$$

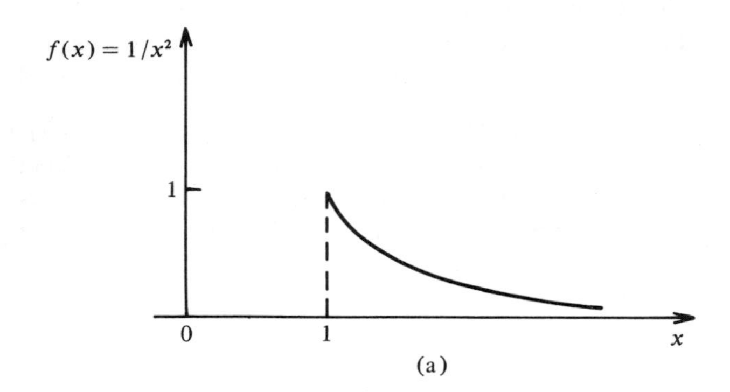


(a)

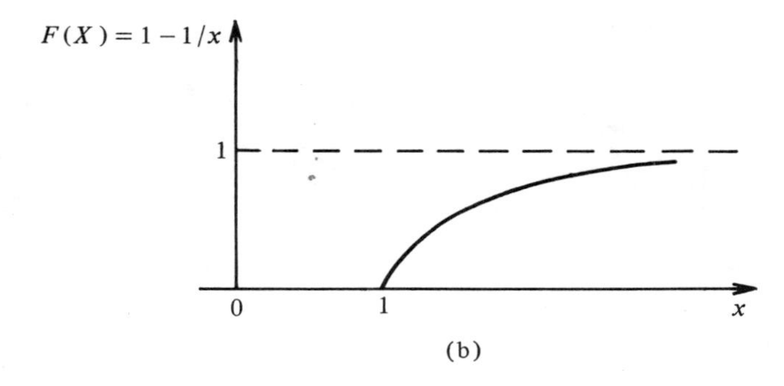


(b)

Fig. 2.8.1

Thus we are unable to set up the equation $U = F(x)$. Other distributions have a nice expression for the CDF $F(x)$, but solving $U = F(x)$ for $X$ can be very difficult. For example the beta distribution with PDF $f(x) = 6x(1-x)$, $0 < x < 1$, can easily be integrated to be $F(x) = 3x^2 - 2x^3$, $0 < x < 1$. But solving $2X^2 - 2X^3 = U$ for $X$ is not simple and quick. Thus, we need other methods to apply in cases in which the inverse CDF method cannot easily be used. These methods will be discussed in the following sections.

## 2.9. Acceptance-Rejection Method of Generating Random Variates

In cases in which the inverse CDF method for continuous distributions cannot be applied, a general technique is available which can be applied so long as the PDF $f(x)$ is known in closed form. The technique is called the (acceptance-) *rejection method of random variate generation*. (It is also known as the *acceptance sampling method*, or the rectangle method.)

Suppose the support $S$ for the distribution is the interval $(a, b)$ along the real line. Let $h = \max_{x \in [a, b]} f(x)$. That is, $h$ is the highest point of the density function between $a$ and $b$. (Note that $h$ could be $f(a)$ or $f(b)$.) Draw a rectangle, completely containing the area under the density function, with horizontal base $(a, b)$ and vertical sides $(0, h)$. (See Fig. 2.9.1.) Notice that the area of the rectangle is $h(b-a) \geqslant 1$, while the area under $f(x)$ is equal to 1.

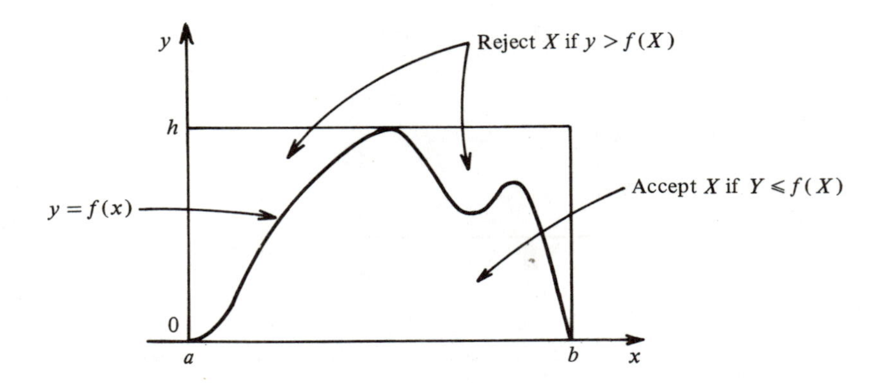


Fig. 2.9.1

The procedure for generating random variates with PDF $f(x)$ follows:

*Algorithm 2.9.1*

1) Let $U_1, U_2$ be two $U(0, 1)$ random numbers.
2) Generate $X = (b-a)U_1 + a$, having a $U(a, b)$ distribution.
3) Generate $Y = hU_2$, having a $U(0, h)$ distribution. Then the point $(X, Y)$ is a random point (uniformly distributed) in the rectangle.
4) Compare $Y$ to $f(X)$. If $Y \leq f(X)$, then ouput $X$ as the desired random number. If $Y > f(X)$, then reject $X$ as unusable.
5) Repeat the process by generating new uniform numbers $U_1$ and $U_2$. With each repetition, a new $X$-value is either accepted (and output), or rejected.

Theoretical justification of Algorithm 2.9.1 requires the background of conditional probability which is beyond the scope of this book. However, an intuitive justification is possible. Since $(X, Y)$'s are uniformly distributed in $(a, b) \times (0, h)$, each point in the rectangle $(a, b) \times (0, h)$ has the same probability to be picked. Thus an acceptable pair $(X, Y)$, that is $Y \leq f(X)$, is in the region bounded by the $x$-axis, the density curve, and the vertical line $x = x_0$. The probability of an acceptable pair $(X, Y)$ is proportional to the area of this region. In other words,

$$\Pr\{X < x_0\} = \text{Area bounded by } x\text{-axis, the density curve, and } x = x_0$$

$$= \int_a^{x_0} f(t)\mathrm{d}t = F(x_0) \quad ,$$

which is exactly what we want.

*Example 2.9.1:* Write a computer program to generate random variates from the density function

$$f(x) = 2310x^4 (1-x)^6 \quad , \quad 0 < x < 1 \quad .$$

*Solution:* It is quite obvious that the inverse CDF method is very difficult to use in this case. To use the acceptance-rejection method, we first need to find a rectangle that contains $f(x)$. By differentiation, we see that

$$f'(x) = 4620x^3 (1-x)^5 (5x-2) \quad .$$

Thus the highest point of $f(x)$ occurs at $x = 2/5$, or $h = f(2/5) = 2.759045$. We use the rectangle $(0, 1) \times (0, 2.76)$ to cover $f(x)$ and generate $X$ by the following algorithm.

1) Generate two random $U(0, 1)$ numbers $U_1$ and $U_2$

2) Let $X = U_1$, $Y = 2.76*U_2$

3) If $Y \leqslant f(X)$ OUTPUT $X$
   Otherwise go to 1.

Notice that the $X$-value is acceptable if and only if the point $(X, Y)$ lies beneath the corresponding point of the curve $(X, f(X))$. If $(X, Y)$ lies above the curve, then $X$ is rejected. This process produces $X$-values with relative frequencies determined by the area of the rectangle. The proportion of $X$-values which will be accepted in the long run is the ratio of the area under $f(x)$ to the total area in the rectangle, i.e., $1/h(b-a)$. This proportion is called the *efficiency* of the procedure. If the proportion is near 1, (that is, the area in the rectangle is only slightly greater than the area under $f(x)$), then the procedure will accept a high proportion of $X$-values, and fewer repetitions will be needed to obtain a given sample size $n$ of $X$-values. On the other hand, if the proportion is near 0, then the procedure is inefficient, since a large number of $X$-values will be unusable. In this case, it would take longer to obtain a sample of $n$ usable $X$-values. In most applications, the rejection method is practical only when a high efficiency is possible.

In cases of low efficiency, we often improve the process by using several rectangles of different heights. For instance, we can divide the support $S$ into several non-overlapping intervals $(a_1, b_1)$, $(a_2, b_2)$, . . . , $(a_k, b_k)$. For each interval, locate $h_i = \max_{x \in [a_i, b_i]} f(x)$, the maximum height of the curve $f(x)$ over the $i$th interval. Draw the $k$ rectangles, completely containing the area under the density function, with horizontal bases $(a_i, b_i)$ and vertical sides $(0, h_i)$. Then the total area in the rectangles is

$$\sum_{i=1}^{k} h_i(b_i - a_i) \geqslant 1 \quad .$$

(See Fig. 2.9.2.) The procedure for generating random numbers is the following. Pick a rectangle at random, using probabilities proportional to the rectangle areas; i.e.,

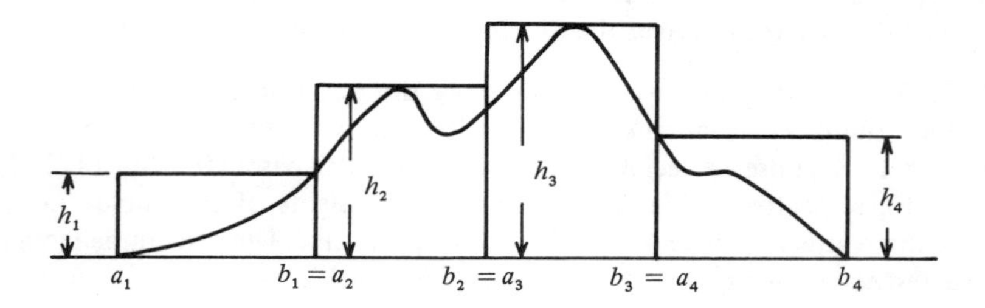


Fig. 2.9.2

$$\Pr(\text{rectangle } i \text{ is picked}) = \frac{\text{area of rectangle } i}{\text{total area of all rectangles}}$$

$$= \frac{h_i(b_i - a_i)}{\sum_{j=1}^{k} h_j(b_j - a_j)} \quad .$$

(What method would you use to do this random selection?) Once the $i$th rectangle has been chosen, generate $X = (b_i - a_i)U + a_i$ and $Y = h_i U$, so that the point $(X, Y)$ is a randomly chosen point (uniformly distributed) inside the rectangle. If $Y \leqslant f(X)$, then output $X$ as the desired random number. Otherwise reject $X$ as unusable, and start from the beginning by picking a rectangle at random. This process should be repeated as long as desired.

As before, the generated $X$-value is acceptable if and only if the point $(X, Y)$ lies beneath the curve $f(x)$. The proportion of $X$-values which will be accepted in the long run is

$$\frac{1}{\text{total area of all rectangles}} = \frac{1}{\sum_{i=1}^{k} h_i(b_i - a_i)} \quad . \tag{2.9.1}$$

The use of several rectangles, rather than just a single rectangle, results in higher efficiencies, that is, more acceptable $X$-values. The disadvantage is that a rectangle must be chosen at each step. Thus, putting in too many rectangles can slow the procedure, by causing a long process in choosing a rectangle at random.

The optimal number of rectangles must then depend on the gain in efficiency from adding rectangles versus the loss from choosing the rectangle at each step.

*Example 2.9.2:* Suppose we wish to apply the rejection technique to generate random numbers from the beta distribution with PDF $f(x) = 6x(1-x)$, $0 < x < 1$. Suppose we decide to use 3 rectangles with bases $(0, 1/3), (1/3, 2/3)$, $(2/3, 1)$, respectively. We find the maximum heights of the curve $f(x)$ over those three intervals to be $4/3, 3/2, 4/3$, respectively. Thus our three rectangles have respective areas of $4/9$, $1/2$, $4/9$. The total area of these is $25/18$. We would choose the rectangle with probabilities

$$\frac{4/9}{25/18} = \frac{8}{25}, \quad \frac{1/2}{15/18} = \frac{9}{25}, \quad \frac{4/9}{25/18} = \frac{8}{25},$$

respectively. Once the rectangle is chosen, $(X, Y)$ is generated within that rectangle, as usual, and the $X$-value is either accepted or rejected.

This method can be easily extended to infinite rectangles provided the total area converges. Otherwise we would have a 0 acceptance rate according to (2.9.1). Take a gamma random variable for example. Let the density function of the gamma random variable $X$ have PDF,

$$f(x) = \frac{1}{4!} x^4 e^{-x}, \quad 0 < x < \infty.$$

Since the CDF of $X$ is a complicated function, the inverse method is difficult to apply. By differentiation, we see that the maximum of $f(x)$ happens at $x = 4$, or the highest point $h = f(4) = 0.19535 \cong 0.2$. Moreover, we know that, by

$$te^{-t} \leqslant e^{-1} \text{ for } t \geqslant 0,$$

$$x^4 e^{-x/2} = (xe^{-x/8})^4$$

$$= 8^4 \left[ \left( \frac{x}{8} \right) e^{-x/8} \right]^4$$

$$\leqslant 8^4 e^{-4} \leqslant 76,$$

we have

$$f(x) \leqslant 76e^{-x/2}/4! \leqslant 3.2e^{-x/2}. \tag{2.9.2}$$

Thus $f(x)$ can be covered by rectangles that cover $g(x) = 3.2e^{-x/2}$ for large $x$. From Fig. 2.9.3, we see that the rectangles $(0.8) \times (0, 0.02)$, $(8, 9) \times (0, g(8))$, $(9, 10) \times (0, g(9))$, ..., $(n, n+1) \times (0, g(n))$, ... cover $f(x)$ and the total area of these rectangles is

$$\begin{aligned} A &= 8 \times 0.2 + g(8) + g(9) + \ldots \\ &= 1.6 + 3.2e^{-4}(1 + e^{-1/2} + e^{-1} + \ldots + e^{-n/2} + \ldots) \\ &= 1.6 + 3.2e^{-4}/(1 - e^{-1/2}) \\ &= 1.6 + 0.477 = 2.077 \quad . \end{aligned}$$

An easier way to get around the method of infinite rectangles is to truncate the $X$ by a large number $M$ such that $\Pr\{|X| \geqslant M\}$ is insignificant. For example, suppose we wish to generate 10,000 random $X$'s and we find that $\Pr\{|X| \geqslant 10\} \leqslant 10^{-8}$. Since there is hardly any chance to generate an $|X| \geqslant M$, we may let the support of $X$ be $(-M, M)$ or $(0, M)$ for non-negative $X$.

An even more general method to generate $X$ is to cover the PDF of $f(x)$ by any function $g(x)$ instead of rectangles, i.e., we find a function $g(x)$ such that $f(x) \leqslant g(x)$ for all $x \in S$, the support of $X$. Moreover we require that $g(x)$ is a much "simpler" function than $f(x)$ and $\int_S g(x)\mathrm{d}x = c < \infty$. Then we can use the following algorithm to generate $X$.

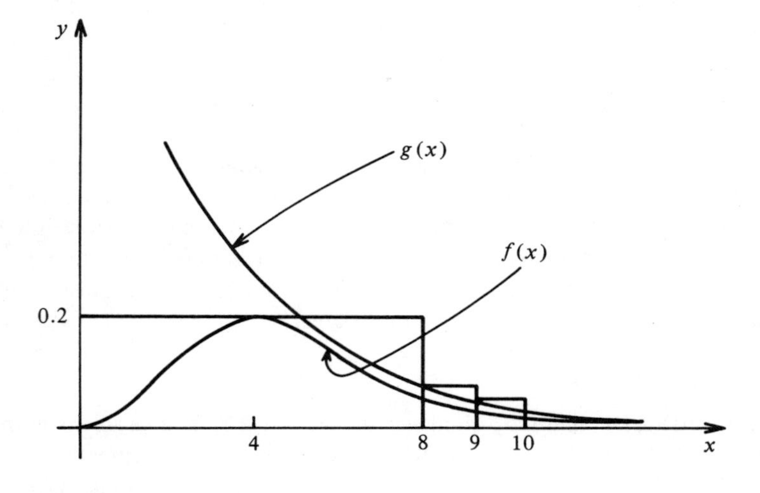


Fig. 2.9.3

*Algorithm 2.9.2*

1) Generate a random variate $X$ according to the density function $h(x) = g(x)/c$.
2) Generate a random number $Y$ uniformly in $(0, g(X))$.
3) If $Y \leqslant f(X)$ output $X$, otherwise GO TO 1.

The intuitive reason for this method can be seen from Fig. 2.9.4. The first two steps of Algorithm 2.9.2 produce an $(X, Y)$ which is uniformly distributed in the area between $g(x)$ and the $x$-axis. The third step can be reasoned similarly to the rectangle method.

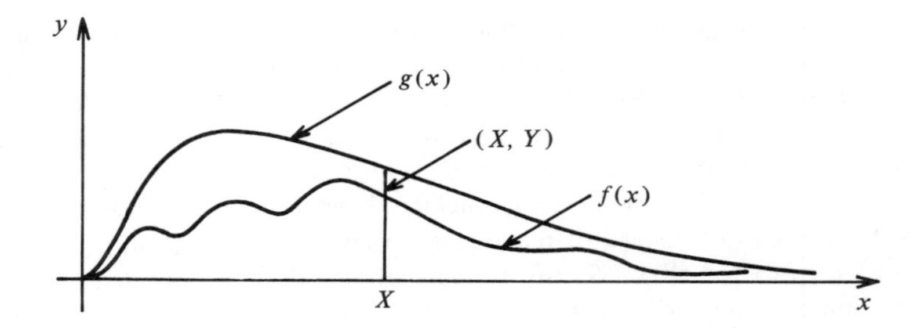


Fig. 2.9.4. Note that $X$ is generated proportionally to the infinitesimal area under $g(x)$ and $Y$ is uniform in $(0, g(X))$, making $(X, Y)$ uniformly distributed under $g(x)$.

## 2.10 Generating Normal Random Variates

The $N(0, 1)$ distribution plays a vital role in almost all areas of statistics, and in particular, in statistical simulations. The need for normal random variates arises quite frequently in such applications, and hence much effort has been devoted to the development of efficient random variate generators for the normal distribution. In this section, we discuss several of the most common methods for generating $N(0, 1)$ random variates.

*1) Inverse CDF method*

The inverse CDF method, discussed in Section 2.8, is a problem with the normal distribution since there is no explicit closed-form expression for the CDF $\Phi(x)$ or its inverse function $\Phi^{-1}(u)$. Approximations have been developed and some of them are quite adequate in most applications. One such approximation is the following:

Generate $U \sim U(0, 1)$.

Define $w = \sqrt{-2 \ln U}$

For $0 < U \leqslant .5$ ,

$$X = \Phi^{-1}(U) = \frac{a_0 + a_1 w + a_2 w^2}{1 + b_1 w + b_2 w^2 + b_3 w^3} - w$$

where

$a_0 = 2.515517$ $\quad a_1 = 0.802853$ $\quad a_2 = 0.0103028$

$b_1 = 1.432788$ $\quad b_2 = 0.189269$ $\quad b_3 = 0.001308$

For $.5 < U < 1$,

$X = -\Phi^{-1}(1 - U)$ using the above approximation.

*2) Rejection methods*

A rejection procedure, as discussed in Section 2.9, can easily be applied to the normal distribution. Such a procedure has been found effective in many random variate generators. One remark on such procedures should be made at this point. Since the support $S$ for the normal distribution is infinite $(-\infty < x < +\infty)$, it may not be practical to cover all of the area under the PDF with finite rectangles. A truncation can be made at each end of the support, placing upper and lower bounds on the values of $X$ which can be generated from such a procedure. Such bounds if chosen sufficiently far out in the tails, usually do not prove to be a serious detriment to the random generator.

As we have seen from Section 2.9 that the tail probability is very important in determining the truncation point, the following result will serve as a guide.

*Theorem* (Feller [8], p. 166). Let $Z \sim N(0, 1)$, then for large $M$, say $M \geqslant 3$,

$$\Pr\{|Z| \geqslant M\} \approx \frac{2}{M\sqrt{2\pi}} e^{-M^2/2} .$$

For example, by choosing $M = 4$, the tail probability

$$\Pr\{|Z| \geqslant 4\} \doteq 0.00007 .$$

*3) Central Limit Theorem*

One of the simplest and widely used methods for generating normal random numbers makes use of the *Central Limit Theorem* (CLT). This famous theorem in statistics tells us the following:

*CLT:* Let $X_1, X_2, \ldots, X_n$ be a random sample of observations from a distribution (discrete or continuous) with mean $\mu$ and variance $\sigma^2$. Let

$$\bar{X} = \frac{1}{n} \sum_{i=1}^{n} X_i \quad .$$

Then for large $n$, $\bar{X}$ is approximately normally distributed with mean $\mu$ and variance $\sigma^2/n$. The approximation improves as $n$ gets larger.

Thus for large $n$, $Z = (\bar{X} - \mu)/(\sigma/\sqrt{n})$ should be approximately distributed as an $N(0, 1)$ random variable. This result gives us a method of generating normal random numbers. Generate $n$ $X$-values from some specified distribution. (It does not matter which distribution is used for the $X$'s, but of course we would prefer one which is easy and fast to do.) Since we know what distribution the $X$-values follow, we can find $\mu$ and $\sigma^2$ for that distribution. Then calculate

$$Z = \frac{\bar{X} - \mu}{\sigma/\sqrt{n}} \quad .$$

This $Z$-value will be an $N(0, 1)$ random variate.

Thus $n$ $X$-values from some distribution are required to generate one $Z$-value. For the sake of efficiency, we would not like to use a large sample size $n$. However, for small $n$, we run the risk of generating $Z$-values which do not closely follow a normal distribution. Thus there is a trade-off between efficiency and accuracy in the algorithm. We must also choose the distribution of the $X$-values for computational simplicity and speed. The most common choice by far is to use the $U(0, 1)$ distribution for the $X$'s. Then $\mu = 1/2$ and $\sigma = 1/\sqrt{12}$ (see Section 2.5), and

$$Z = \frac{\bar{X} - 1/2}{1/\sqrt{12n}} \sim N(0, 1) \quad .$$

We can further improve the efficiency of the algorithm if we can eliminate the square root from the calculation. This is done by choosing $n = 12$, so that we get

$$Z = \frac{\bar{X} - 1/2}{1/12} = 12\bar{X} - 6 = \sum_{i=1}^{12} X_i - 6 \quad .$$

This greatly simplifies the calculations. Some IBM computers use this method to generate $N(0, 1)$ random variates.

*4) Box-Muller Method I*

Another method for generating normal random variates is due to Box, Muller, and Marsaglia. The procedure is as follows:

Generate $U_1$ and $U_2$, two $U(0, 1)$ random variates.

Compute $V_1 = 2U_1 - 1$ and $V_2 = 2U_2 - 1$ ,

and $S = V_1^2 + V_2^2$ .

If $S \geqslant 1$, then generate new values for $U_1$, $U_2$ and recompute $V_1$, $V_2$, and $S$.

If $S < 1$, compute $$Z_1 = V_1 \sqrt{\frac{-2 \ln(S)}{S}} \quad ,$$

$$Z_2 = V_2 \sqrt{\frac{-2 \ln(S)}{S}} \quad .$$

Then $Z_1$, $Z_2$ are two $N(0, 1)$ random variates. Thus pairs of uniform random numbers are required to produce pairs of standard normal random numbers. Notice also that certain pairs of uniforms must be discarded, which lessens the efficiency of the algorithm. It can be shown that

$$\Pr(S \geqslant 1) = 1 - \pi/4 \cong 0.2146 \quad .$$

Thus about 21% of the pairs of generated uniforms will not be usable.

The advantage of the Box-Muller method is that it is *exact*: the random variates produced using this method will really have a standard normal distribution, not just an approximation. However, it is more complicated to program and is found to be relatively time-consuming in large studies.

*5) Box-Muller Polar Method*

A second method due to Box and Muller (1958) is the polar method, which again uses a pair of $U(0, 1)$ random numbers to generate a pair of $N(0, 1)$ $Z$-values. The procedure is:

Generate $U_1$ and $U_2$, two uniform random numbers.

Compute $Z_1 = \sqrt{-2 \ln U_1} \; \cos(2\pi U_2)$ ,

$$Z_2 = \sqrt{-2 \ln U_1} \quad \sin(2\pi U_2) \quad .$$

$Z_1$ and $Z_2$ are a pair of $N(0, 1)$ random variates.

The polar method is exact, just as Box-Muller Method I is. Also this method is generally considered superior in terms of speed and efficiency to Method I. (Note that the polar method does not reject any uniform pairs.)

**Exercise 2**

2.1 Find the CDF $F(x)$ for the random variable which is the outcome of a fair die toss. Find also the mean and the variance of this random variable.

2.2 Prove (2.2.1). (Hint: By mathematical induction and Ex. 1.10. $\Pr\{X = x\} = p\Pr\{$there are $x - 1$ successes in the first $n - 1$ trials$\} + (1 - p)\Pr\{$ there are $n$ successes in the first $n - 1$ trials$\}$.)

2.3 (a) Show that the binomial probabilities defined in (2.2.1) can be computed by the following recursive formula.

(i) $p_0 = q^n$, $q = 1 - p$

(ii) $p_x = [(n - x + 1)p/(xq)]\, p_{x-1}$, $x = 1, 2, \ldots, n$.

(b) Let $X \sim \mathrm{Bi}(n, p)$ and $np = \lambda$. Show that for large $n$ and fixed small $p$,

$$\Pr\{X = x\} \cong e^{-x}\lambda^x/x!\ , \qquad x = 0, 1, \ldots$$

(Hint: Use the fact $(1 - \lambda/n)^n \to e^{-\lambda}$, as $n \to \infty$.)

2.4 (*Programming exercise*) Do a computer simulation to check the validity of the binomial formula $p_i = \binom{n}{i} p^i q^{n-i}$. Compare the theoretical values with the simulation values by 1,000 simulations. (Let $n = 5$, $p = 0.43$.)

2.5 We know that $X \sim \mathrm{Bi}(n, p)$ can be approximated by a Poisson distribution for large $n$ and small $p$ (see Ex. 2.3). Find the binomial and Poisson probabilities in the following table. The numbers in the parentheses are $(n, p, X)$.

| | (10, 0.1, 0) | (10, 0.1, 1) | (20, .05, 2) |
|---|---|---|---|
| By Binomial | ________ | ________ | ________ |
| By Poisson | ________ | ________ | ________ |

2.6 A book of 500 pages contains 500 misprints. Assume that the misprints are randomly distributed in the book,

(a) find the probability that page 241 has no misprints, and

(b) find the probability that there are no pages that contain more than 3 misprints from pages 241 to 250, inclusive.

2.7 The defect rate of a particular brand of computer chips is $\alpha$, and a computer is made of 500 of this type of chips. Find the maximum allowable defect rate such that more than 95% of the computers built will work after being assembled.

2.8 The lifetime of an electronic component is normally distributed with mean 1,000 hours and standard deviation 200 hours. Let $X_1, X_2, \ldots, X_5$ be the lifetimes of 5 randomly chosen components. Find

(i) $\Pr\{X_1 > 800\}$,

(ii) $\Pr\{800 < X_1 < 1250\}$,

(iii) $\Pr\{|X_1 - 1000| > 100\}$,

(iv) $\Pr\{X_1 + X_2 + \ldots + X_5 > 4000\}$,

(v) $\Pr\{\min(X_1, X_2, \ldots, X_5) > 800\}$.

2.9 A rope is made of 50 threads. It is known that each thread has a mean breaking strength of 0.8 lb. with standard deviation 0.3 lb. Suppose the breaking strength of the rope is the sum of the breaking strengths of its components. Find the probability that this rope has breaking strength less than 35 lbs. (Hint: Use the central limit theorem and the fact $\Sigma X_i = n\overline{X}$.)

2.10 In the previous exercise, what is the minimum number of threads required in a rope in order to guarantee that 99% of the ropes have breaking strength more than 40 lbs.?

2.11 Let $Z$ be the standard normal random variable. Find the following probabilities. $\Pr\{-1.2 < Z < 2.1\}$, $\Pr\{Z > 1.41\}$, $\Pr\{|Z| \leqslant 1.3\}$, and $\Pr\{|Z - 1| \leqslant 0.8\}$. Also find the $x$'s so that the following probability statements are correct. $\Pr\{Z \leqslant x\} = 0.90$, $\Pr\{|Z| > x\} = 0.10$, $\Pr\{|Z| \geqslant x\} = 0.05$.

2.12 Find the expected value of the exponential distribution defined in (2.4.4).

2.13 Suppose the expected waiting time for a telephone call in a secretary's office is 2 minutes. Find the probability that she has to wait more than three minutes for her first call.

2.14 In the previous problem, what is the probability that she will have $X$ calls in a 3-minute interval if $X$ is considered as a Poisson distribution. Compare this answer with the answer in Ex. 2.13 and make a general statement for them.

2.15 (*Programming exercise*) Construct a statistical distribution library that contains the standard normal, $t$, chi-square, and $F$ probabilities. Check your output with Table 2.6.1.

2.16 In generating the random variates defined by Table 2.7.1, two methods can be used. Which one of the following is faster? Can you set a general rule for fast discrete random variate generations?

| *Algorithm 1* | *Algorithm 2* |
|---|---|
| Let $U \sim U(0, 1)$ | Let $U \sim U(0, 1)$ |
| If ( $U \leqslant 0.25$ ) $X = 0$ Output | If ( $U \leqslant 0.5$ ) $X = 1$ Output |
| If ( $U \leqslant 0.75$ ) $X = 1$ Output | If ( $U \leqslant 0.75$ ) $X = 0$ Output |
| $X = 2$ Output | $X = 2$ Output |

For Problems 2.17-2.20, assume your computer takes one unit of time unit of time to do addition, subtraction, and comparison, 7 units of time to do multiplication, 32 units of time to do division, 83 units of time to do elementary functions such as $\sqrt{\ }$, log, sin, cos, and exp, and 10 units of time to generate a random number.

2.17 Compute the average times required to generate one $X$ by the algorithms in Ex. 2.16.

2.18 Compare the average time required by the following four methods in generating standard normal variates.

(i) Inverse CDF,
(ii) the Central Limit Theorem,
(iii) the acceptance-rejection method by a rectangle $(-4, 4) \times (0, 0.4)$ as cover, and
(iv) Box and Muller's polar method.

2.19 Find an efficient way to generate a random variate from each of the following density functions.

(i) $f(x) = 2x, \ 0 < x < 1$
(ii) $f(x) = 0.5e^{-0.5x}, \quad x > 0$
(iii) $f(x) = 12x^2(1-x), \quad 0 < x < 1.$

2.20 Write down an algorithm to generate a Poisson random variate and compute the average time to generate it.

2.21 (*Programming exercise*) Write down a histogram subroutine that will produce $c$ (input) equal class interval histogram. The upper and lower bounds for the histogram can be provided either by the user as inputs or to be determined by the highest and lowest data. This subroutine will be used to do the following.

(a) Take 1,000 random numbers, $X_i = U_i \sim U(0, 1)$, $i = 1, 2, \ldots, n$ ($n = 1{,}000$) and plot their histograms with $c = 10$ in $(0, 1)$.
(b) Do (a) for $X_i = U_i^{1/2}$, $i = 1, 2, \ldots, n$. (Use $c = 20$ in $(0, 1)$.)
(c) Do (a) for $X_i = -\ln(U_i)$, $i = 1, 2, \ldots, n$. (Use $c = 20$, in $(0, 10)$.)

(d) Plot the histograms for $\overline{X} \equiv$ the average of 10 consecutive $X$'s. (There are 100 $\overline{X}$'s. Use $c = 10$, and the suitable upper and lower bounds for the histogram so that they reveal the shape of the distribution of $\overline{X}$.)

2.22 (*Programming exercise*) Generate 100 random variates from the following populations and plot their histograms.

(i) Binomial Bi(10, 0.7)

(ii) Poisson ($\lambda = 5$)

(iii) Normal $N(2.0, 4.0)$

(iv) $X$ with density function

$$f(x) = 12x^2(1-x), \quad 0 < x < 1.$$

Chapter 3

# THE BINOMIAL DISTRIBUTION AND ITS APPLICATIONS TO SIMULATION

## 3.1. Introduction

So far we have learned how to use a computer for random sampling from a given population, and we know that by taking repeated samples, we get information about the variation between samples (see Chapter 1). But how many samples are necessary in order to draw a useful conclusion? Recall Table 1.2.1, for example. The probability statement (1.2.1)

$$\Pr\{|\hat{\theta} - \theta| \geqslant 0.1\} \cong .005$$

is based on 1,000 samples. But is this enough (or too many, perhaps) for statement (1.2.1) to be accurate? The number 1,000 seems to have been picked either by intuition or by convenience. We will waste our energy (and computing time) if we do too many simulations. On the other hand, we may draw an erroneous conclusion if we do not do enough simulations. The art of simulation size determination will be discussed in this chapter.

If we go back to Table 1.2.1, we will naturally ask about the reliability of statement (1.2.1) when we know it was based on 1,000 simulated samples. It can easily be seen that the reliability of (1.2.1) would be increased if it had been based on 10,000 simulations. To assess the reliability of (1.2.1), we can superimpose another simulation experiment, but the reliability of the super-imposed experiment has to be determined by another super-superimposed simulation. Hence we see that the reliability of (1.2.1) cannot be determined by simulation alone. Fortunately, we can investigate this problem using probability theory involving only elementary mathematics.

Let's formulate our problem in an abstract manner. Let $p$ be the probability that a statement A is true. Suppose we have found that in $n$ trials, the statement A was true $r$ times. What can we say about the probability that statement A is true? Consider (1.2.1), for example. Our statement A is:

$$|\hat{\theta} - \theta| \geqslant 0.1 \quad .$$

Our $p$ is:

$$p = \Pr\{|\hat{\theta} - \theta| \geqslant 0.1\} = \Pr\{\text{A is true}\} \quad .$$

From Table 1.2.1 of Chapter 1, we have $n = 1{,}000$ and $r = 5$. Our question is, can we say that $p \cong 0.005$?

To further simplify the notation, we state the problem as follows:

Let $p$ be true probability of success on each trial.

Suppose $n$ independent trials have been tried and $r$ successes have been obtained. We wish to know the relationship between $\hat{p} = r/n$ and the true unknown $p$.

*Note:* When we say that trials are independent, we mean that the results of the previous trials will not influence the outcome of the succeeding trials.

We will start to assess the relationship between $\hat{p}$ and $p$ in the next section.

## 3.2. The Binomial Distribution and Its Normal Approximation

We have seen in Chapter 2 that the probability of observing $X = r$ successes in $n$ independent trials, when the probability of success is $p$, is

$$\Pr(X = r) = \binom{n}{r} p^r (1-p)^{n-r} \quad \text{for } r = 0, 1, 2, \ldots, n \ , \qquad (3.2.1)$$

where

$$\binom{n}{r} = \frac{n!}{r!(n-r)!}$$

is the number of different combinations when $r$ elements are selected from $n$ different elements. We say that $X$ has a binomial distribution and denote it by $X \sim \mathrm{Bi}(n, p)$. For small $n$, a calculator can easily compute (3.2.1) for any given $p$ and $r$, but for large $n$, say $n > 200$, even a computer has difficulty storing numbers such as $n!$, $r!$, or $(n-r)!$. Unfortunately, we are usually interested in a large number of trials. It has been shown (Feller, Vol. I), that, for large $n$, (3.2.1) can be approximated by a normal distribution. More precisely, if $X \sim \mathrm{Bi}(n, p)$, then for large $n$,

$$\Pr(X=i) \cong \Pr\left\{i-\tfrac{1}{2} < Y < i+\tfrac{1}{2}\right\} , \tag{3.2.2}$$

where

$$Y \sim N(np, npq), \qquad \text{with} \quad q = 1-p \quad .$$

By standardizing $Y$, (3.2.2) can be reduced to

$$\Pr\{X=i\} = \Pr\left\{\frac{i-\frac{1}{2}-np}{\sqrt{npq}} < Z < \frac{i+\frac{1}{2}-np}{\sqrt{npq}}\right\}$$

$$= \Phi\left(\frac{i+\frac{1}{2}-np}{\sqrt{npq}}\right) - \Phi\left(\frac{i-\frac{1}{2}-np}{\sqrt{npq}}\right) \quad . \tag{3.2.3}$$

Figure 3.1.1 and Table 3.2.1 demonstrate the goodness of the normal approximation, for even a moderate $n (n=10,\ p=.4)$. More examples are in the exercises. It has been shown that the approximation (3.2.3) is very good if $npq > 9$, and reasonably good if $n \geqslant 20$ and $p$ is not too close to the extreme values 0 or 1.

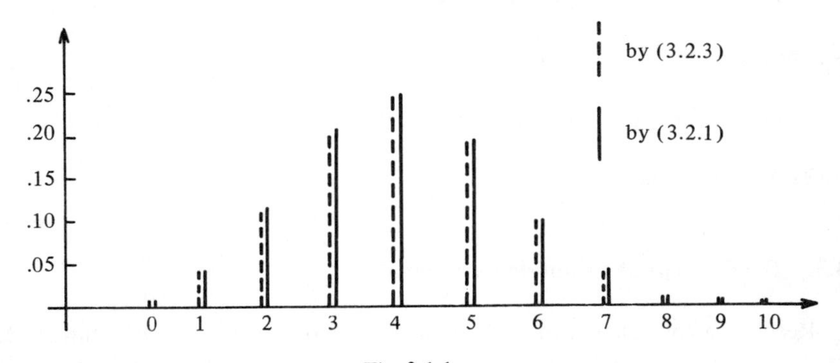


Fig. 3.1.1

Table 3.2.1

| | $\Pr(X = i)$ using | |
|---|---|---|
| $i$ | (3.2.1) | (3.2.3) |
| 0 | .006 | .012 |
| 1 | .040 | .041 |
| 2 | .121 | .113 |
| 3 | .215 | .207 |
| 4 | .251 | .253 |
| 5 | .201 | .207 |
| 6 | .111 | .113 |
| 7 | .042 | .041 |
| 8 | .011 | .010 |
| 9 | .002 | .002 |
| 10 | .000 | .000 |

Formula (3.2.2) is sometimes written as

$$\Pr\{X = i\} \cong \Pr\left\{i - \tfrac{1}{2} < X < i + \tfrac{1}{2}\right\}, \quad \text{for } X \sim N(np, npq)$$

for convenience. And the *continuity correction factor* $\frac{1}{2}$ is usually dropped if we are interested in a large range of $X$ values. For example,

$$\Pr\{i \leq X \leq j\} \quad \text{for} \quad X \sim \text{Bi}(n, p)$$

is approximately

$$\Pr\{i \leq X \leq j\} \quad \text{for} \quad X \sim N(np, npq)$$

for large $n$ and large $|j - i|$.

## 3.3. The Concept of a Confidence Interval

Recall that our goal is to find the relationship between $\hat{p} = r/n$ and $p$. Let us assume the sample size $n$ is large. (For small sample sizes, see Section 3.7.)

Suppose we have not yet observed $r$. Then we denote $r$ by the random variable $X$, and our estimate $\hat{p}$ by

$$\hat{p} = X/n \quad .$$

By (3.2.2), we know that

$$n\hat{p} = X \sim N(np, npq),$$

approximately, or

$$Z = \frac{n\hat{p} - np}{\sqrt{npq}} \sim N(0, 1)$$

or

$$Z = \frac{\hat{p} - p}{\sqrt{pq/n}} \sim N(0, 1) \quad . \tag{3.3.1}$$

From (3.3.1), we may use the normal tables and make the following probability statement, i.e.,

$$\Pr\left\{\left|\frac{\hat{p} - p}{\sqrt{pq/\sqrt{n}}}\right| < z_{1-\alpha/2}\right\} = 1 - \alpha \quad . \tag{3.3.2}$$

We may interpret (3.3.2) in a different way. That is, (3.3.2) tells us, with probability $(1 - \alpha)$,

$$-z_{1-\alpha/2} \leqslant \frac{\hat{p} - p}{\sqrt{pq/n}} \leqslant z_{1-\alpha/2} \quad . \tag{3.3.3}$$

Solving (3.3.3) for $p$, we have with probability $(1 - \alpha)$,

$$\frac{\hat{p} + (Z^2/2n)}{1 + (Z^2/n)} - D \leqslant p \leqslant \frac{\hat{p} + (Z^2/2n)}{1 + (Z^2/n)} + D \quad , \tag{3.3.4}$$

where

$$D = \frac{Z}{1 + (Z^2/n)} \left\{\frac{\hat{p}(1 - \hat{p})}{n} + \frac{Z^2}{4n^2}\right\}^{1/2} \quad ,$$

and $Z = z_{1-\alpha/2}$. The interval obtained in (3.3.2) is usually called the $(1-\alpha)$ confidence interval for $p$. Sometimes, we say that we have $(1-\alpha)$ 100% confidence that $p$ lies between the two values expressed in (3.3.4). No matter how it is stated, (3.3.4) means that if we construct the interval

$$\left(\frac{\hat{p}+(Z^2/2n)}{1+(Z^2/n)} - D\,,\ \frac{\hat{p}+(Z^2/2n)}{1+(Z^2/n)} + D\right)$$

100 times, it will cover the true $p$ approximately $(1-\alpha)100$ times.

When $n$ is large and $p$ is not too small, we may let

$$1+\frac{Z^2}{n} \simeq 1\,, \quad \hat{p}+\frac{Z^2}{2n} \simeq \hat{p}\,, \quad \text{and } D \simeq Z\sqrt{\frac{\hat{p}(1-\hat{p})}{n}}\,.$$

Thus a simpler $1-\alpha$ confidence interval for $p$ is

$$\left(\hat{p} - z_{1-\alpha/2}\sqrt{\frac{\hat{p}(1-\hat{p})}{n}}\,,\ \hat{p} + z_{1-\alpha/2}\sqrt{\frac{\hat{p}(1-\hat{p})}{n}}\right). \tag{3.3.5}$$

We can now make a formal statement about

$$p = \Pr\{|\hat{\theta}-\theta| \geqslant 0.1\} \tag{3.3.6}$$

from Chapter 1. By simulation, we found $\hat{p} = 0.005$ with $n = 1{,}000$. If we desire a 95% confidence interval for $p$, then by (3.3.4) the interval is

(0.0021, 0.0117)

and by (3.3.5), the confidence interval is

(0.0006, 0.0094).

The 99% confidence intervals are, respectively,

(0.0017, 0.0149)

and (0, .0107).

Note that the method developed in this section is not only good for making probability statements based on simulation results, but also good for estimating a proportion based on a sample from a general population. We will apply this technique in the next section to a different type of simulation problem.

## 3.4. Monte Carlo Methods

In general, Monte Carlo methods may be defined as any procedures which use statistical sampling to obtain approximate solutions to mathematical or physical problems. Monte Carlo, of course, is the resort commune in Monaco, on the Mediterranean, famous for its gambling casinos. The Monte Carlo method is so named because there is an element of gambling involved — the approximate solution we obtain is not like that in the usual numerical analysis sense, but is instead a *stochastic* (probabilistic) approximation.

Consider the following general problem: Suppose that $a$ is some geometrical figure. (For graphical purposes, we shall assume it to be a two-dimensional region.) We are interested in the area of $a$: Area $(a) = A$. Suppose that $a$ is completely contained in some rectangle $\beta$, and that Area $(\beta) = B$ is known. See Fig. 3.4.1.

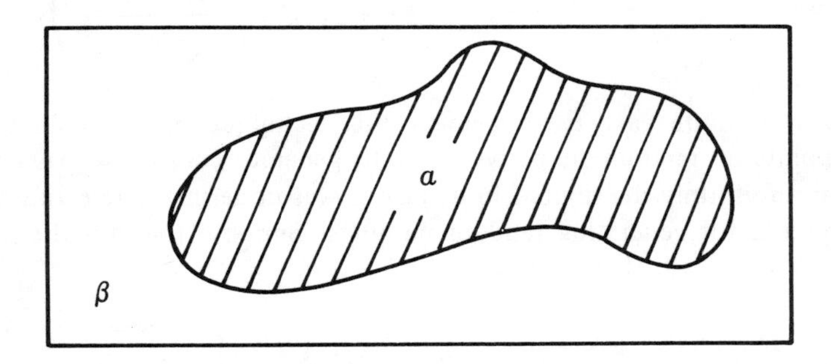


Fig. 3.4.1

Now consider an experiment in which points are selected at random from $\beta$. What is the probability that a randomly selected point $(x, y)$ belongs to $a$? From the elementary rules of probability we have that

$$\Pr\{(x, y) \in a\} = \frac{\text{Area } (a)}{\text{Area } (\beta)} = \frac{A}{B} .$$

Suppose then that $n$ points are selected (sampled) independently and randomly from $\beta$. Let $X$ be the random variable which denotes the number of these $n$ points which also belong to $a$. Clearly, $X$ has a binomial distribution with parameters $n$ and $p = A/B$. Thus we can estimate

$$p = \frac{A}{B} \text{ by } \hat{p} = \frac{X}{n} .$$

Since $B$ is known, we can then estimate $A$ by $B \cdot \hat{p}$. Furthermore, when $n$ is sufficiently large, the results of the last sections (or instead, the Central Limit Theorem) imply that $B\hat{p}$ will be approximately normally distributed with mean $= A$ and variance $= A(B - A)/n$; i.e.,

$$\hat{A} = B\hat{p} \sim N\left(A, \frac{A(B-A)}{n}\right) .$$

Thus we can form a confidence interval for $A$ by multiplying through in either (3.3.4) or (3.3.5) by the constant $B$. Using (3.3.5), for example, we obtain as a $1 - \alpha$ confidence interval for $A$:

$$\left(B\hat{p} - z_{1-\alpha/2}\sqrt{\frac{\hat{A}(B-\hat{A})}{n}}\ ,\quad B\hat{p} + z_{1-\alpha/2}\sqrt{\frac{\hat{A}(B-\hat{A})}{n}}\right). \qquad (3.4.1)$$

Now it is quite easy to implement this technique on the computer. To select points at random in $\beta$, we need to generate coordinate pairs $(X, Y)$ which are uniformly distributed in $a$, just as was done in the rejection method of Chapter 2 for generating random numbers. Suppose that $\beta$ is the rectangle $(x_1, x_2) \times (y_1, y_2)$; that is

$$\beta = \{(x, y) : x_1 \leqslant x \leqslant x_2 , \quad \text{and} \quad y_1 \leqslant y \leqslant y_2\} .$$

Recall that if $U_1$ is a uniform random number in (0, 1), then $(x_2 - x_1)U_1 + x_1 = X$ is a uniform random number in $(x_1, x_2)$. Similarly, if $U_2$ is distributed as $U(0, 1)$, then $(y_2 - y_1)U_2 + y_1 = Y$ is a uniform random number in $(y_1, y_2)$. Then the point $(X, Y)$ is a random point uniformly distributed in $\beta$.

We now summarize the procedure:

*Algorithm 3.4.1*

1) Generate $2n$ independent uniform random numbers in (0, 1) and arrange them in $n$ pairs.

2) Use each pair $(U_1, U_2)$ to generate a corresponding point $(X, Y)$ in $\beta$.

3) Check whether or not the point $(X, Y)$ is in $a$. Usually the boundary of $a$ is given by an equation that can easily be programmed.

*Example:* Suppose we wish to find the area inside the ellipse defined by

$$\frac{x^2}{4} + \frac{y^2}{9} = 1 \quad .$$

(See Fig. 3.4.2.)

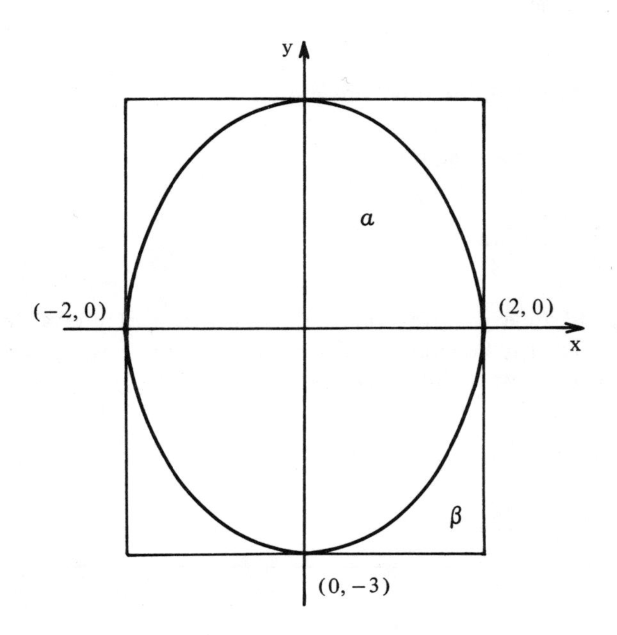


Fig. 3.4.2

A sample of 1,000 points in $\beta$ were picked at random and 774 of these were found to be inside $\alpha$ (i.e., $X = 774$). Then our estimate of $A$ where $B = 24$ (the area of $\beta$) is

$$\hat{A} = B \cdot \frac{774}{1{,}000}$$

$$= 18.576 \quad .$$

A 95% confidence interval for $A$, according to (3.4.1) is given by

$$\left(\hat{A} - z_{0.975}\sqrt{\frac{\hat{A}(B-\hat{A})}{1000}}\;,\quad \hat{A} + z_{0.975}\sqrt{\frac{\hat{A}(B-\hat{A})}{1000}}\right)$$

$$= (17.954,\ 19.198)\quad .$$

It turns out that the true value of $A = 6\pi = 18.850$. Thus, this particular interval did, in fact, cover the true value of $A$.

## 3.5. Monte Carlo "Hit or Miss" Integration

Suppose that we are required to compute

$$A = \int_a^b f(x)\,\mathrm{d}x \quad ,$$

where $f(x)$ is some positive-valued function over the interval $(a, b)$. We will assume that $f(x)$ is bounded on $(a, b)$; that is, there is a positive number $M$ such that $|f(x)| \leqslant M$ for all $x \in (a, b)$. (See Fig. 3.5.1.)

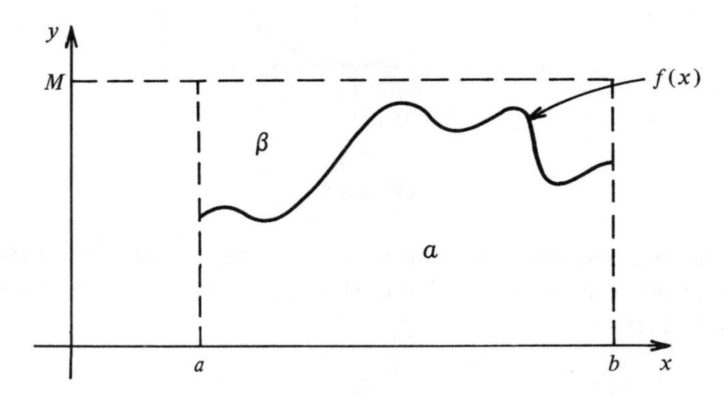


Fig. 3.5.1

Then $A$ is the area under the curve $f(x)$, over $(a, b)$. Thus the region $\alpha$ is the area defined by the graph of the function $f(x)$, the $x$-axis, and $x = a$ and

$x = b$. Since $|f(x)| \leqslant M$, we can take $\beta$ to be the rectangle

$$\beta = \{(x, y): a \leqslant x \leqslant b., \ 0 \leqslant y \leqslant M\} \ .$$

We can now apply the previous results to obtain an estimate and confidence interval for $A$.

*Example 3.5.1:* Suppose we want to find $A = \int_0^1 1/\sqrt{2\pi}\, e^{-x^2/2}\, dx$. (Do you recognize this $f(x)$?) See Fig. 3.5.2.

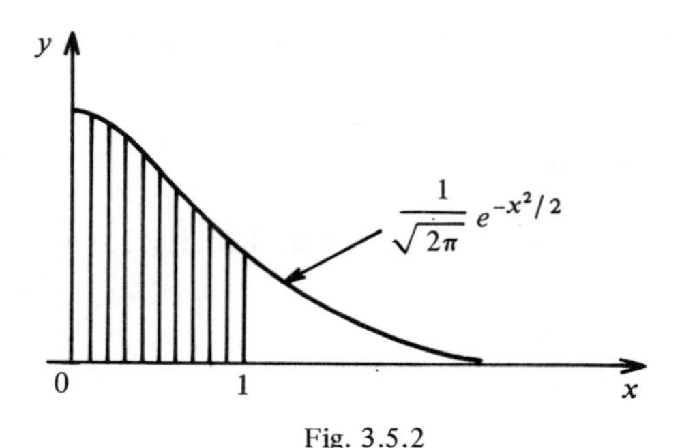


Fig. 3.5.2

Note that $f(x)$ is strictly decreasing over $(0, 1)$. Thus we can take $M = f(0) = 1/\sqrt{2\pi}$. Using the computer, then, we would sample points $(x, y)$ in $\beta = (0, 1) \times (0, 1/\sqrt{2\pi})$. We determine for each point, whether or not $y \leqslant f(x)$.

We implemented this procedure on a computer, using $n = 10{,}000$, and found that $X = 8589$. Our estimate of $A$ is then

$$A = B \cdot \hat{p} = \frac{1}{\sqrt{2\pi}}(0.8589) = .3427 \ .$$

A 99% confidence interval for $A$ is, using (3.4.1),

$$\left( .3427 - 2.57 \sqrt{\frac{.3427(.3989 - .3427)}{10{,}000}} \ , \right.$$

$$\left. .3427 + 2.57 \sqrt{\frac{.3427(.3989 - .3427)}{10{,}000}} \right)$$

$$= (.3390, .3462) \ .$$

In this case, we can find the true value of $A$ from the normal tables. It turns out that $A = .3413$. Once again, our particular interval does cover the true value of $A$ in this case.

Of course, we would resort to Monte Carlo "hit or miss" integration only in a case where the $f(x)$ function is too complicated to solve the problem analytically (i.e., to use the methods of calculus to directly perform the integration) Thus Monte Carlo integration is an alternative to numerical integration techniques, such as Simpson's rule. The obvious disadvantage in using Monte Carlo is that we do not obtain a definite solution — instead we obtain a confidence interval. However if we are to choose $n$ large enough, we can make the width of the confidence interval as small as we would like. On the other hand, the advantage of using the Monte Carlo technique is its ease to implement.

### 3.6. Sample Size Determination

Interval (3.3.5) may be restated as, with probability $1 - \alpha$

$$|\hat{p} - p| \leq z_{1-\alpha/2}\sqrt{\frac{\hat{p}\hat{q}}{n}} \quad . \qquad (3.6.1)$$

Suppose we plan to choose a sample of size $n$ (unknown), and wish to have $(1 - \alpha)$ 100% confidence that

$$|\hat{p} - p| < \epsilon \quad , \qquad (3.6.2)$$

where $\alpha$ and $\epsilon$ are given. To solve for the required sample size $n$, we equate (3.6.1) and (3.6.2). Thus

$$z_{1-\alpha/2}\sqrt{\frac{\hat{p}\hat{q}}{n}} = \epsilon$$

or

$$n = \frac{z^2_{1-\alpha/2}\,\hat{p}\hat{q}}{\epsilon^2} \quad . \qquad (3.6.3)$$

Formula (3.6.3) cannot be applied, since we have not yet observed $\hat{p}$ and $\hat{q}$. But a result from mathematics shows

$$\hat{p}\hat{q} = \hat{p}(1 - \hat{p}) \leq 1/4 \quad .$$

Thus

$$n \leqslant \frac{z_{1-\alpha/2}^2}{4\epsilon^2} \quad . \tag{3.6.4}$$

That is, if we choose the sample size as large as the right-hand side of (3.6.4), then (3.6.2) is guaranteed no matter what value of $p$ is observed.

We can also apply this result to the Monte Carlo integration technique of the last section. There, if we want

$$|\hat{A} - A| < \epsilon \tag{3.6.5}$$

and we proceed as in (3.6.3) we obtain

$$n = z_{1-\alpha/2}^2 \frac{\hat{A}(B - \hat{A})}{\epsilon^2} \quad .$$

Again, since $\hat{A}$ is unknown prior to taking the sample, we can be conservative if we replace $\hat{A}(B - \hat{A})$ by its upper bound $B^2/4$. Then we obtain

$$n \leqslant \frac{z_{1-\alpha/2}^2 B^2}{4\epsilon^2} \quad . \tag{3.6.6}$$

Using this upper bound for $n$ guarantees that we will have $100(1-\alpha)\%$ confidence that $|\hat{A} - A| \leqslant \epsilon$.

It should be noted that the needed sample size increases as $B$ increases. Therefore it is wise to find the rectangle $\beta$ which encloses $a$ as tightly as possible.

*Example 3.6.1:* Suppose we want to approximate

$$A = \int_0^1 \frac{x(1-x)}{(1-x/2)^2} \, dx \quad .$$

What value of $n$ would enable us to obtain an estimate of $A$ that would, with 99% confidence, be within .001 of the true value?

We find that $f(x)$ has a maximum value of $\frac{1}{2}$ at $x = \frac{2}{3}$ (using calculus, or numerically). Thus we take $\beta = (0, 1) \times (0, \frac{1}{2})$. According to (3.6.6), we should choose

$$n = \frac{(2.57)^2 (\frac{1}{2})^2}{4(.001)^2} = 412{,}806 \quad .$$

This may or may not be feasible, according to available funds!

## 3.7. The Concept of Hypothesis Testing

Suppose we wish to find the relationship between $\hat{p}$ and $p$ for small $n$. We know in this case the normal approximation cannot be used and consequently we do not have a nice probability statement such as (3.3.2) to describe the relationship between $\hat{p}$ and $p$. But we still may be able to make an inference about $p$ under a different formulation.

For example, suppose we have taken a random sample of 10 people and none of them have tuberculosis. We wish to make an inference about the true rate of tuberculosis in the population. The normal approximation to the binomial is obviously not applicable, due to the small sample size, nor are the formulae (3.3.4) and (3.3.5), which were derived from this approximation. But we may still be able to make a statement about the true tuberculosis rate, such as it cannot be as high as 0.9. We will reason such a conclusion this way.

Suppose the tuberculosis rate is 0.9 or higher. Then the probability of observing 0 cases of tuberculosis out of 10 people is

$$\begin{aligned} P &= \binom{10}{0} p^0 (1-p)^{10} \qquad (p \geqslant 0.9) \\ &= (1-p)^{10} \\ &\leqslant (1-0.9)^{10} = 0.000000001 \quad . \end{aligned} \tag{3.7.1}$$

Since this probability is so small, we would say that it is very unlikely that $p \geqslant 0.9$. On the other hand, can we conclude that the rate of tuberculosis is less than 0.1? (This seems to be a conservative conclusion, since our estimate $p = 0/10 = 0.0$.)

We will use the same kind of reasoning. If the rate of tuberculosis is no less than 0.1, what is the probability that we will get 0 cases from the 10 people? The probability is

$$\begin{aligned} p &= \binom{10}{0} p^0 (1-p)^{10} \qquad (p \geqslant 0.1) \\ &= (1-p)^{10} \\ &\leqslant (1-0.1)^{10} = 0.349 \quad . \end{aligned} \tag{3.7.2}$$

Thus we see that the probability can be as high as 0.349 when $p \geqslant 0.1$. Consequently, it is probably too risky to claim $p \leqslant 0.1$.

The present approach is different from the one of confidence intervals because we do not estimate $p$ (which is difficult). We simply assess the likelihood of whether a given hypothesis is true or not. To state this approach formally, we usually need two hypotheses called $H_0$ and $H_1$, as well as a reasonable decision rule, and the amount of risk we can afford when a false $H_1$ is accepted. Consider the tuberculosis example for instance. The first set of hypotheses of interest were

$$H_0 : p \geqslant 0.9$$
$$H_1 : p < 0.9 \quad . \tag{3.7.3}$$

The decision rule is to accept $H_1$ if we see no tuberculosis cases in a sample of size 10, and our probability of accepting $H_1$ when it is false is

$$\Pr\{X = 0 \mid H_1 \text{ is false}\} \ , \tag{3.7.4}$$

where $X$ is the number of TB cases in a sample of size $n = 10$, and the vertical bar is followed by the conditions under which we compute the probability.

Formula (3.7.4) is equivalent to

$$\Pr\{X = 0 \mid H_0 \text{ is true}\} = \Pr\{X = 0 \mid p \geqslant 0.9\}$$
$$\leqslant 0.0000000001 \quad . \tag{3.7.5}$$

Thus if we accept $H_1$, the likelihood that we have made a mistake is less than 0.0000000001.

The second set of hypotheses of interest are

$$H_0 : p \geqslant 0.1$$
$$H_1 : p < 0.1 \quad . \tag{3.7.6}$$

The decision rule is the same as before and the probability of accepting a false $H_1$ is

$$\Pr\{X = 0 \mid H_1 \text{ is false}\}$$
$$= \Pr\{X = 0 \mid H_0 \text{ is true}\}$$
$$= \Pr\{X = 0 \mid p \geqslant 0.1\}$$
$$= (1 - p)^{10} \quad . \qquad (p \geqslant 0.1) \tag{3.7.7}$$

As we have seen from (3.7.2), the above probability can be as high as 0.349. Thus it is quite risky to accept $H_1$.

Let us summarize the three basic steps in hypothesis testing:

1) State two hypotheses $H_0$ and $H_1$.
2) Find a reasonable decision rule for when to accept $H_1$.
3) Compute $\alpha = \Pr\{\text{Accepting } H_1 \text{ under this rule} \mid H_1 \text{ is false}\}$. This $\alpha$-value will be referred to as the *risk* or the *significance level* of the test. We call the decision error a *Type I error*. (3.7.8)

A few comments are necessary on hypothesis testing:

(a) Usually, $H_1$ is accepted according to a decision rule for which $\alpha$, defined in (3) above, is quite small (such as .05). Thus, when $H_1$ is accepted, it is likely to be true. Consequently, we should always make $H_1$ to be the hypothesis which we would like to "prove" or accept with a high degree of confidence.

(b) In (2) above, we wish not only to find a reasonable decision rule for accepting $H_1$, but we would like it to be the "best" decision rule possible for accepting $H_1$. The derivation of the best decision rules are usually quite complicated mathematical problems. But for simple statistical problems, the best decision rules usually agree with the reasonable or intuitive decision rules.

(c) When we fail to accept $H_1$, it does not mean that $H_0$ is necessarily true. It is possible that $H_1$ is true, and yet our test fails to show it. The probability that this occurs is

$$\beta = \Pr\{\text{Not accepting } H_1 \text{ under the decision rule} \mid H_1 \text{ is true}\}.$$

(We call this type of decision error a *Type II error*.) In later sections, we will be interested in a quantity called the *power* of a test:

$$\begin{aligned} \text{Power} &= 1 - \beta \\ &= \Pr\{\text{Accepting } H_1 \mid H_1 \text{ is true}\} . \end{aligned}$$

Naturally, we wish our test to have small $\beta$ (near 0) and high power (near 1). Unfortunately, we will see that this is not always possible.

(d) The hypotheses in (3.7.3) and (3.7.6) are usually called *composite* hypotheses, in contrast to *simple* hypotheses where $H_0$ and $H_1$ each consists of only one value of $p$. For example, with

$$H_0 : p = 0.1$$

$$H_1 : p = 0.5 \quad , \tag{3.7.9}$$

both hypotheses are simple ones. For a simple hypothesis, the risk (probability of accepting a wrong $H_1$) and the power (probability of accepting a correct $H_1$) are each single values under a given decision rule. But this is not true for composite hypotheses (consisting of more than one value of $p$), as we have seen in (3.7.5) and (3.7.7). We can only find a bound for the risk or power under composite hypotheses. Thus, the risk $\alpha$ is defined to be the *maximum possible risk* under all possible values in the composite hypothesis $H_0$. It usually happens at the boundary point of $H_0$. For example $H_0$ is $p \geqslant 0.9$ in (3.7.3) and the maximum possible risk happens at $p = 0.9$.

(e) If we review (3.7.6)

$$H_0 : p \geqslant 0.1$$

$$H_1 : p < 0.1 \tag{3.7.10}$$

again, we see that we would have a likely chance of making a wrong decision if $p = 0.0999999$. Such a value of $p$ belongs to $H_1$, but is very close to $H_0$. In this case, the power would certainly be very low. Indeed the power would be very close to $\alpha$, since $p$ is very close to $H_0$. Thus we are usually interested in the power of this test only when the $p$ in $H_1$ is not so close to $H_0$. For example, in testing (3.7.10), we might ask the power of the test only for $p \leqslant .05$. Thus we define the power of a test for a composite $H_1$ to be the minimum power (guaranteed power) under all possible values in a subset of $H_1$.

In other words, we want the following information concerning our decision rule:

(1) What is the risk of accepting a false $H_1$?

(2) What is the guaranteed power of accepting a correct $H_1$ when for example, $p \leqslant 0.05$?

As in the maximum risk, the minimum (guaranteed) power also happens in most cases at the boundary.

*Example 3.7.1:* Let $p$ be the unknown defect rate of a large shipment of electronic components. We can accept this shipment as satisfactory if $p$ is less than 5%. A sample of 100 components will be examined, and we decide to accept this shipment if at most one component is defective. What is our risk of accepting a bad lot (defined as $p \geqslant 0.05$), and what is our power in accepting this lot if $p \leqslant 0.01$?

*Solution:* Hypotheses:

$$H_0 : p \geqslant 0.05$$

$$H_1 : p < 0.05$$

*Decision rule:* Accept $H_1$ if 1 or less components are defective in a sample of 100.

$$\text{Risk} = \Pr\{0 \text{ or } 1 \text{ components are defective} \mid H_0 \text{ is true}\}$$

$$= \binom{100}{0} p^0 (1-p)^{100} + \binom{100}{1} p^1 (1-p)^{99}, \text{ where } p \geqslant .05$$

$$= (1-p)^{100} + 100p(1-p)^{99}$$

$$\leqslant (1-.05)^{100} + 100(.05)(1-.05)^{99} = .037 \quad .$$

$$\text{Power} = \Pr\{0 \text{ or } 1 \text{ components are defective} \mid p \leqslant .01\}$$

$$= (1-p)^{100} + 100p(1-p)^{99} \qquad \text{where } p \leqslant .01$$

$$\geqslant (1-.01)^{100} + 100(.01)(1-.01)^{99}$$

$$= .736 \quad .$$

Thus the maximum risk in this test is .037 and the guaranteed power for $p \leqslant .01$ is 0.736.

## 3.8. Determination of Acceptance Rules and Sample Sizes in Hypothesis Testing

Let us begin this section with Example (3.7.1) in Section 3.7. In that example, our decision rule was to accept $H_1$ if one or fewer components are defective. There is no reason to say that this is the best decision rule. First we have to admit that our decision rule should depend on the number of defectives in the sample, and secondly, our decision rule must be *monotonic*; that is, if we decide to accept $H_1$ when there is one defective in the sample, we should also accept $H_1$ when there are no defectives in the sample. A non-monotonic decision rule would be, for example, accepting $H_1$ if there is one defective, but not accepting $H_1$ if there are no defectives. Such a rule is obviously not optimal and will not be considered here. Thus our decision rules are of the form: "accept $H_1$ whenever there are $c$ or fewer defectives in the sample." Table 3.8.1 shows the risk (significance level) and power for different $c$ values (decision rules).

Table 3.8.1 ($n = 100$)

| | | $c$ | | | | | |
|---|---|---|---|---|---|---|---|
| | | 0 | 1 | 2 | 3 | 4 | 5 |
| Risk | | .006 | .037 | .118 | .265 | .440 | .616 |
| Power | $p \leqslant 0.03$ | .048 | .195 | .420 | .647 | .815 | .916 |
| | $p \leqslant 0.01$ | .366 | .736 | .945 | .981 | .996 | .999 |

We see that the merit of any decision rule is represented by a trade-off between risk and power. If we wish to have a smaller risk, we must sacrifice some power. If we want to decrease the risk and simultaneously increase the power, we have to increase the sample size. Here are the risks and powers under some larger sample sizes (see Tables 3.8.2, 3.8.3).

Table 3.8.2 ($n = 200$)

| $c =$ | | 1 | 2 | 3 | 4 | 5 |
|---|---|---|---|---|---|---|
| Risk | | .000 | .003 | .010 | .030 | .055 |
| Power | $p \leqslant 0.03$ | .017 | .062 | .151 | .285 | .446 |
| | $p \leqslant 0.01$ | .406 | .677 | .858 | .947 | .983 |

Table 3.8.3 ($n = 500$)

| | $c =$ | 10 | 12 | 14 | 16 | 18 |
|---|---|---|---|---|---|---|
| | Risk | .001 | .005 | .016 | .40 | .092 |
| Power | $p \leqslant 0.03$ | .118 | .268 | .466 | .664 | .819 |
| | $p \leqslant 0.01$ | .986 | .995 | .998 | .999 | 1.000 |

Thus, we see that the best decision rule, the risk, the power at a certain subset of $H_1$, and the sample size are all related. Usually the desired risk and power are given, and the sample size and best decision rule will be determined by examining tables similar to Tables 3.8.1-3.8.3. Sometimes the sample size is fixed for economical or other reasons. In that case we may have to sacrifice either the risk or the power.

For large sample hypothesis testing problems involving the binomial distribution, analytic formulas can be used to determine the sample size and the acceptance region (critical region) of the decision rule. (See Table 3.8.4.)

Table 3.8.4

| Hypotheses | Risk | Power | Sample Size |
|---|---|---|---|
| $\begin{cases} H_0 : p \geqslant p_0 \\ H_1 : p < p_0 \end{cases}$ | $\alpha$ | $1-\beta$ at $p \leqslant p_1$ | $n = \left( \dfrac{z_{1-\beta}\sqrt{p_1 q_1} - z_\alpha \sqrt{p_0 q_0}}{p_0 - p_1} \right)^2$ |
| $\begin{cases} H_0 : p \leqslant p_0 \\ H_1 : p > p_0 \end{cases}$ | $\alpha$ | $1-\beta$ at $p \geqslant p_1$ | $n = \left( \dfrac{z_{1-\alpha}\sqrt{p_0 q_0} - z_\beta \sqrt{p_1 q_1}}{p_1 - p_0} \right)^2$ |

The derivation of Table 3.8.4 is not at all difficult. Suppose the hypotheses are $H_0 : p \geqslant p_0$ against $H_1 : p < p_0$. A reasonable decision rule is obviously $X \leqslant c$, where $X$ is the number of successes in $n$ trials when $p$ is defined to be the probability of success in each trial. Since the risk $\alpha$ happens at the boundary, $p = p_0$,

$$\alpha = \Pr\{X \leqslant c \mid p = p_0\}$$

$$= \Pr\left\{ Z \leqslant \frac{c - np_0}{\sqrt{np_0 q_0}} \right\} , \qquad (3.8.1)$$

where $q_0 = 1 - p_0$ and the transfer from $X$ to $Z$ is by the normal approximation to $\text{Bi}(np_0, np_0q_0)$. Obviously (3.8.1) is equivalent to

$$\frac{c - np_0}{\sqrt{np_0q_0}} = z_\alpha \quad . \tag{3.8.2}$$

Similarly the minimum power happens at the boundary $p_1$. Thus,

$$1 - \beta = \Pr\{X \leqslant c \mid p = p_1\} \quad ,$$

$$= \Pr\left\{Z \leqslant \frac{c - np_1}{\sqrt{np_1q_1}}\right\}$$

or

$$\frac{c - np_1}{\sqrt{np_1q_1}} = z_{1-\beta} \quad . \tag{3.8.3}$$

Eliminating $c$ in (3.8.2) and (3.8.3) and solving for $n$, we have the sample size in Table 3.8.4.

*Example 3.8.1:* In the quality control scheme discussed in Example 3.7.1, find the sample size and decision rule if we would like our probability of accepting $H_1 : p < 0.05$ when actually $p \geqslant 0.05$ be 0.05 and the probability of getting the shipment accepted is 0.90 when $p \leqslant 0.03$.

*Solution:* By the notation of Table 3.8.4, we have $p_0 = 0.05$, $p_1 = 0.03$, $\alpha = 0.05$, $1 - \beta = 0.9$, or $\beta = 0.1$. Thus $z_\alpha = z_{0.05} = -1.645$, $z_{1-\beta} = z_{0.9} = 1.282$,

$$n = \left(\frac{1.282\sqrt{.03 \times .97} + 1.645\sqrt{0.05 \times 0.95}}{0.05 - 0.03}\right)^2$$

$$= \left(\frac{0.219 + 0.359}{0.02}\right)^2$$

$$= 834 \quad ,$$

and the decision rule is to choose, according to (3.8.2),

$$\frac{c - np_0}{\sqrt{np_0q_0}} = z_\alpha \quad ,$$

or

$$c = np_0 + z_\alpha \sqrt{np_0 q_0}$$

$$= 834 \times 0.05 + (-1.645)\sqrt{834 \times 0.05 \times 0.95}$$

$$= 41.7 - 10.35$$

$$= 31.35$$

$$\simeq 31 \quad .$$

Thus our decision rule is to take a sample of size $n = 834$ and accept $H_1$ if and only if the number of defectives $r \leqslant 31$. By doing this we guarantee the required risk and power.

### 3.9. Two-sided Tests and the Relationship Between Hypotheses Testing and Confidence Intervals

We have seen from the previous section that the hypotheses on $p$, the probability of success in a binomial distribution, are one-sided hypotheses, i.e., $H_0$ takes values on one end of the interval [0, 1] and $H_1$ takes all other values (the other end). Sometimes, however, we are interested in testing whether $p$ is different from a given value. For example, if we wish to test whether a coin is balanced or not, we need to test the hypotheses

$$H_0 : p = 0.5$$

$$H_1 : p \neq 0.5 \quad . \tag{3.9.1}$$

The hypothesis $H_1$ in (3.9.1) is called a two-sided hypothesis, and the test to make decisions about $H_0$ and $H_1$ based on data is called a two-sided test. The principles involved in testing the hypotheses (3.9.1) are the same as those discussed in the previous section. The three steps are, according to (3.7.8):

1) State the hypotheses, which in general form, are

$$H_0 : p = p_0$$

$$H_1 : p \neq p_0$$

where $p_0$ is any given probability of success or a given proportion.

2) The decision rule is to accept $H_1$ if

$$|X - np_0| \geqslant c \quad , \tag{3.9.2}$$

where $X$ is the number of successes observed in $n$ trials and $c$ is an integer to be determined by the choice of the risk (significance level) that will be used.

3) The risk is given by $\alpha = \Pr\{|X - np_0| \geqslant c \mid H_0\}$ .

*Example 3.9.1:* We wish to check whether a coin is unbalanced by tossing it one hundred times. Determine the acceptance region when the risk is set at $\alpha = 0.05$.

*Solution:* Since the sample size is fixed, there is no need to use the power in determining the acceptance region. Moreover, the normal approximation to the binomial distribution can safely be applied, since the criterion $npq > 9$ is satisfied with $n = 100$ and $p = 0.5$. Letting $p$ = probability of observing a head on any one toss, and the hypotheses

$$H_0 : p = 0.5$$

$$H_1 : p \neq 0.5 \quad ,$$

we need to find $c$ such that

$$\begin{aligned}\alpha &= 0.05 \\ &= \Pr\{|X - 100 \times 0.5| \geqslant c \mid X \sim \mathrm{Bi}(100, 0.5)\} \\ &= \Pr\{|X - 50| \geqslant c \mid X \sim N(50, 25)\} \\ &= \Pr\{|Z| \geqslant c/5\} \quad .\end{aligned} \tag{3.9.3}$$

Thus, we must have $c/5 = z_{1-\alpha/2} = 1.96$ or $c \simeq 10$. That is, we should accept $H_1$ with risk (significance level) 0.05 whenever we observe $|X - 50| \geqslant 10$, i.e., $X \geqslant 60$ or $X \leqslant 40$. Suppose we tossed the coin 100 times and the head appeared 62 times. Then we can conclude that the coin is not balanced at 0.05 level of significance.

Suppose that, based on this result (62 heads), we wish to use the confidence interval to make an inference on $p$. What will our conclusion be? By the formula introduced in (3.3.5), a 95% confidence interval for $p$ is

$$\left(\hat{p} - 1.96\sqrt{\frac{\hat{p}\hat{q}}{n}}\ , \qquad \hat{p} + 1.96\sqrt{\frac{\hat{p}\hat{q}}{n}}\right)$$

$$= (0.62 - (1.96)(0.049), \quad 0.62 + (1.96)(0.049))$$

$$= (0.52, 0.72) \quad .$$

Thus we have 95% confidence that $p \geqslant 0.52 > 0.5$, which is the same conclusion we derived from the hypothesis test. Actually, one can easily show the following fact:

For large $npq$ (or $npq$), we accept $H_1$ in the test of

$$H_0 : p = p_0$$

$$H_1 : p \neq p_0$$

at the $\alpha$ level of significance if and only if $p_0$ is not covered by a $(1-\alpha)$ confidence interval for $p$.

*Example 3.9.1* (continued): Suppose we tossed the coin 100 times, and 42 heads were obtained. What would be our inference on $p$ at the $\alpha = 0.05$ level of significance (i.e., with 95% confidence)?

*Solution:* We may do it by hypothesis testing or confidence interval:

1) By the hypothesis testing criterion, we do not accept $H_1 : p \neq 0.5$ at $\alpha = 0.05$.

2) A 95% confidence interval for $p$ is

$$0.42 \pm (1.96)\sqrt{\frac{(0.42)(0.58)}{100}} = (0.32, 0.52) \quad ,$$

which covers 0.5.

## 3.10. The $p$-value

When a statistical finding is reported, most people do not like the statement "The hypothesis $H_1$ is accepted at 0.05 level of significance." Such a statement does not tell the reader anything about whether $H_1$ can still be accepted at the 0.01 level, or perhaps the 0.001 level. Perhaps the reader feels that a risk of 0.05 is too high for his own particular purposes, and he can accept this hypothesis only when the risk is 0.01 or lower. Thus there is a tendency in the statistical community to report the lowest risk level that the alternative hypothesis can still be accepted. This lowest risk level is usually referred to as the $p$-value, or the observed significance level. For example, if we see a report saying

"Drug A is more effective than drug B (at $p = 0.001$),"

this means that the alternative hypothesis,

$H_1$ : Drug A is more effective than drug B,

can be accepted at any risk $\geqslant 0.001$, but cannot be accepted at any risk level smaller than 0.001.

The difference between this $p$-value and the significance level $\alpha$ is that $\alpha$ is determined before taking any observations, while $p$-value is found after the data have been collected and analyzed. The $\alpha$-level is still commonly used in sample size determination when no data are available, but at the report stage we usually report the $p$-value.

*Example 3.10.1:* A coin is tossed 100 times and 63 heads are observed. What can we say about the hypothesis that this coin is unbalanced?

*Solution:* Let us look at the conclusion we reach under different $\alpha$-levels when testing

$$H_0 : p = 0.5 \quad \text{vs.}$$

$$H_1 : p \neq 0.5 \quad ,$$

where $p$ is the probability of getting a head in each toss. From (3.9.3) we know that the decision to accept $H_1$ should be made when $|X - 50| \geqslant 5z_{1-\alpha/2}$ at $\alpha$ level of significance.

| $\alpha$ | 0.05 | 0.04 | 0.03 | 0.02 | 0.01 | 0.005 | 0.002 |
|---|---|---|---|---|---|---|---|
| $C = 5(z_{1-\alpha/2})$ | 10 | 10 | 11 | 12 | 13 | 14 | 15 |
| Conclusion | Accept $H_1$ | Accept $H_1$ | Accept $H_1$ | Accept $H_1$ | Accept $H_1$ | Do not Accept $H_1$ | Do not Accept $H_1$ |

Thus the $p$-value for accepting $H_1$ must be somewhere between 0.01 and 0.005. Finding the exact value is quite easy, since we know that we accept $H_1$ if and only if

$$|63 - 50| \geqslant 5 \cdot z_{1-\alpha/2}$$

or

$$z_{1-\alpha/2} = 2.6 \quad .$$

Thus the exact $p$-value is, from the Z-table, 0.0094.

Another way to look at the $p$-value for this example is as follows: We have two hypotheses $H_0 : p = 0.5$ and $H_1 : p \neq 0.5$ and our decision rule is $|\hat{p} - 0.5n| > c$ for some $c$. Note here $\hat{p}$ is the general estimate of $p$ from a sample *before* being collected. Hence $\hat{p}$ is a random variable and we can talk about its distribution. The risk $\alpha$ or $c$ can be determined by

$$\alpha = \Pr\{|\hat{p} - 0.5| > c \,|\, H_0\} \tag{3.10.1}$$

when one of them is given. When the data has been observed, the $\hat{p}$ estimated from this particular data is no longer a random variable, it is a value. Let us denote the present estimate $\hat{p}$ by $\hat{p}_e$. Obviously, according to (3.10.1) we can accept $H_1$ if and only if we let $c \leqslant |\hat{p}_e - 0.5|$. Thus the lowest risk that we can accept $H_1$ is

$$p\text{-value} = \Pr\{|\hat{p} - 0.5| \geqslant |\hat{p}_e - 0.5|\} \quad . \tag{3.10.2}$$

Here we emphasize again that $\hat{p}$ is a random variable but $\hat{p}_e$ is a numerical quantity. Using normal approximation that $\hat{p} \sim N(0.5,\ 0.25/n)$ under $H_0$ to (3.10.2), we have

$$\begin{aligned} p\text{-value} &= \Pr\left\{|Z| \geqslant \frac{|\hat{p}_e - 0.5|}{0.5/\sqrt{n}}\right\} \\ &= \Pr\left\{|Z| \geqslant \frac{|0.63 - 0.5|}{0.5/\sqrt{100}}\right\} \\ &= \Pr\{|Z| \geqslant 2.6\} = 0.0094 \quad . \end{aligned}$$

## 3.11. Application to Simulation

In this section, we will demonstrate how statistical inference can be made by simulation.

*Example 3.11.1:* A sample of size 10 is taken from a population with unknown mean $\mu$ and unknown variance $\sigma^2$. If the sample mean $\overline{Y}$ is used to make an inference about the mean $\mu$, can we determine the value of $p$ in the following statement

$$\Pr\{|\overline{Y} - \mu| \leqslant 0.1\} = p? \tag{3.11.1}$$

*Solution:* Suppose we made up a reasonable guess of a distribution with mean $\mu$ and the variance $\sigma^2$ and draw 10 pieces of data from it. Then the chance that $|\overline{Y} - \mu| \leqslant 0.1$ is, by definition, the unknown value $p$. If we simulate a sample of $\overline{Y}_1, \overline{Y}_2, \ldots, \overline{Y}_{n_s}$ from this population $n_s$ times, and find that $|\overline{Y} - \mu| < 0.1$ occurs $r$ times, then our estimate of $p$ is $\hat{p} = r/n_s$, and the error of our estimate at, say 99% confidence is

$$\hat{p} \pm 2.576 \sqrt{\frac{\hat{p}\hat{q}}{n_s}} \quad .$$

Suppose we require the accuracy of our estimate of $p$ to be within 0.05 at the 99% confidence level. Then, according to (3.6.4), the number of simulations should be

$$n_s = \frac{2.576^2}{4(0.05)^2} = 664 \quad .$$

We can thus simulate any specific distribution that we desire. Of course, the value of $p$ depends on what distribution we decide to sample our 10 observations from.

Now, this problem can be solved in a different way, if we are willing to assume that we are sampling from a *normal* population, $N(\mu, \sigma^2)$ for some unknown $\mu$ and $\sigma^2$. From Chapter 2, we know that when the sample size is $n = 10$,

$$\overline{Y} - \mu \sim N(0, \sigma^2/10) \quad .$$

Thus (3.11.1) is equivalent to

$$p = \Pr\{|W| \leqslant 0.1\} \quad , \tag{3.11.2}$$

where $W \sim N(0, \sigma^2/10)$, if we sample from a normal population. Since $\mu$ does not appear in this last expression, we do not need to know $\mu$ in order to evaluate $p$. But solving the problem with the unknown $\sigma^2$ requires further insight into the normal distribution. First, we must realize that the $p$ in (3.11.1) does depend on the value of $\sigma^2$. The larger the population variance, the harder it is to get an accurate estimate of $\mu$, thus yielding a smaller value of $p$. There is no way to get around the problem of the dependence of $p$ on $\sigma^2$, unless we alter the probability statement (3.11.1) to

$$\Pr\{|\bar{Y}-\mu| \leqslant 0.1S\} = p \quad , \qquad (3.11.3)$$

where $S$ is the sample standard deviation (see 2.5.1). Equation (3.11.3) still gives us a confidence interval for $\mu$, since $\bar{Y}$ and $S$ are computed from the sample. That is, with confidence $p$, we have

$$\bar{Y} - 0.1S \leqslant \mu \leqslant \bar{Y} + 0.1S \quad .$$

Estimation of $p$ in (3.11.3) can be done easily by simulation, without any knowledge of $\mu$ and $\sigma^2$. The reason for this is explained in the following arguments.

To make things a little more general, we let the sample size be $n$ instead of 10, and the bound be $\epsilon$ instead of 0.1. Thus we wish to find $p$ defined by

$$\Pr\{|\bar{Y}-\mu| \leqslant \epsilon S\} = p \qquad (3.11.4)$$

or equivalently,

$$\Pr\left\{\frac{|\bar{Y}-\mu|}{S} \leqslant \epsilon\right\} = p \quad .$$

Note that $Y_1, Y_2, \ldots, Y_n$ come in from an $N(\mu, \sigma^2)$ population. If we let

$$Z_i = \frac{Y_i - \mu}{\sigma} \quad ,$$

then $Z_1, Z_2, \ldots, Z_n$ come from an $N(0, 1)$ population. By simple algebra, we find

$$Y_i = \mu + \sigma Z_i \quad , \qquad \bar{Y} = \mu + \sigma\bar{Z} \quad ,$$

$$S^2 = \sigma^2 \sum_{i=1}^{n} \frac{(Z_i - \bar{Z})^2}{(n-1)} \quad ,$$

$$\frac{\bar{Y}-\mu}{S} = \frac{\sigma\bar{Z}}{\sigma\sqrt{\sum_{i=1}^{n}(Z_i-\bar{Z})^2/(n-1)}} = \frac{\bar{Z}}{S_Z}$$

where

$$S_Z^2 = \text{sample variance for } Z$$

$$= \sum_{i=1}^{n} \frac{(Z_i - \bar{Z})^2}{(n-1)} .$$

Thus, we see (3.11.4) is completely equivalent to

$$p = \Pr\left\{ \frac{|\bar{Z}|}{S_Z} \leqslant \epsilon \right\} ,$$

where the sample $Z_1, Z_2, \ldots, Z_n$ come from a standard normal distribution. Now the simulation can be done easily because the population for $Z$ is completely defined. Table 3.11.1 gives some simulation results for $p$, based on $n_s = 664$ samples.

Table 3.11.1. Values of $p$ for (3.11.4).

| | | $\epsilon$ | | | |
|---|---|---|---|---|---|
| | | 0.2 | 0.1 | 0.05 | 0.001 |
| $n$ | 10 | 0.45 | 0.23 | 0.12 | 0.02 |
| | 20 | 0.60 | 0.34 | 0.15 | 0.02 |
| | 30 | 0.72 | 0.40 | 0.20 | 0.04 |
| | 40 | 0.80 | 0.47 | 0.24 | 0.05 |

Actually the value $p$ can be found theoretically using a distribution called the $t$-distribution (see Sections 2.6 and 5.2). Its mathematical form is quite complicated, but when it was originally discovered by Gosset around 1900, it was done by hand simulation. The discovery of the $t$-distribution is a milestone in the history of statistics.

More precisely, if $Y_1, Y_2, \ldots, Y_n$ came from a normal distribution with unknown mean $\mu$ and unknown variance $\sigma^2$, then

$$T = \sqrt{n}\,(\bar{Y} - \mu)/S \qquad (3.11.5)$$

has a $t$-distribution with $\nu = n - 1$ degrees of freedom. A table for some percentiles of $t$-distribution is given in Table A.2.

It can also be shown that for large $n$, say $n \geqslant 30$, the $T$ in (3.11.5) has approximately a standard normal distribution and it is not sensitive to the normality assumption of $Y$. This result, sometimes referred to as the central limit theorem with estimated variance, is extremely useful in stimulation because $n$ is usually large.

*Example 3.11.2:* Suppose cars can be parked along a 150-foot long street where the parking spaces are undivided. Suppose, for simplicity, each car has to occupy a space of 15 feet (including the car length and the leeway to get in and out). Since there is no dividing marks on the street, cars will park randomly within the space that is available. Find the mean number of cars $\mu$ that can park there when all the available spaces are used.

*Solution:* To solve this problem by probability theory does not seem easy. Let us do a simulation study. Without loss of generality we may let the street have length 1 and car length 0.1. We first observe the following 3 facts:

1) If a gap between two cars (or street ends) is less than 0.1, no car can be parked in between.
2) If a gap between two cars (or street ends) is between 0.1 and 0.2, one and only one car can park there.
3) If the gap is greater than 0.2, then a car will park in it at a random location within the gap and produces two new gaps.

Doing 1), 2), 3) repeatedly by a computer, we can find the number of cars on this street when all the available space is occupied. We have done a simulation of $n = 200$ times and our record is like 7, 7, 8, 8, 9, 6, 9, . . . The mean is $\bar{Y} = 7.22$ and the sample variance $S^2 = 0.020$. Thus, our estimate for $\mu$ is 7.22 and a 95% confidence interval for $\mu$ is by

$$\Pr\left\{\frac{\sqrt{n}\,|\bar{Y} - \mu|}{S} \leqslant 1.96\right\} = 0.95 \quad ,$$

equals to $7.22 \pm 1.96\sqrt{0.02/200} = 7.22 \pm 0.02$.

*Example 3.11.3:* An electrical company wants to estimate the proportion of trees that are taller than $T_0 = 60$ feet. They will take a sample of $n = 200$ trees and measure their heights, say, $Y_1, Y_2, \ldots, Y_n$. Their estimate of $p$ will be the usual sample proportion, i.e.,

$$\hat{\theta} = r/n \quad ,$$

where $r =$ the number of $Y$'s greater than 60. One day, a person with some knowledge of statistics comes up with a different idea. He argues that since the heights of trees are normally distributed, they may use $Y_1, Y_2, \ldots, Y_n$ to estimate the mean and variance of the forest population. Then the formula

$$\begin{aligned} \theta &= \Pr\{Y > T_0\} \\ &= \Pr\left\{Z > \frac{T_0 - \mu}{\sigma}\right\} = 1 - \Phi\left(\frac{T_0 - \mu}{\sigma}\right) \end{aligned} \tag{3.11.6}$$

suggests that $\theta$ can be estimated by

$$\tilde{\theta} = 1 - \Phi\left(\frac{T_0 - \overline{Y}}{S}\right) \quad , \tag{3.11.7}$$

where $\overline{Y}$ and $S$ are the sample mean and standard deviation from $Y_1, Y_2, \ldots, Y_n$. The question is whether or not $\tilde{\theta}$ is better than $\theta$.

*Solution:* Suppose $Y_i \sim N(\mu, \sigma^2)$. Let $Z_i = (Y_i - \mu)/\sigma$. Then $\theta = r/n$, $r =$ number of $Z_i$'s $\geqslant (T_0 - \mu)/\sigma \equiv T'_0$, and

$$\tilde{\theta} = 1 - \Phi\left(\frac{T'_0 - \overline{Z}}{S_Z}\right)$$

where $S_Z$ is the sample standard deviation of $Z_1, Z_2, \ldots, Z_n$.

Since $T'_0 = (T_0 - \mu)/\sigma$ involves two unknown parameters, $\mu$ and $\sigma$, we can conclude that $\tilde{\theta}$ is better than $\hat{\theta}$ only if it is better over a large range of values of $T'_0$. Since each $Z_i \sim N(0, 1)$ it is quite easy to collect samples $Z_1, Z_2, \ldots, Z_n$ by simulation. To determine the number of samples we need, we must define our goal for the simulation. Let $\theta$ be the real proportion (unknown) defined in (3.11.6). Let

$$p = \Pr\{|\tilde{\theta} - \theta| < |\hat{\theta} - \theta|\} \ ,$$

i.e., the proportion of times that $\tilde{\theta}$ is closer to the true $\theta$. Thus, we are testing the hypotheses

$$H_0 : p \leqslant 0.5$$

$$H_1 : p > 0.5 \quad .$$

Suppose we set $\alpha = 0.01$, $\beta = 0.05$ when $p \geqslant 0.7$. Then the sample size can be computed by Table 3.8.4 as

$$n_s = \left( \frac{2.326\sqrt{0.5 \times 0.5} + 1.645\sqrt{0.7 \times 0.3}}{0.7 - 0.5} \right)^2 \simeq 92 \quad .$$

The following results were obtained (see Table 3.11.2).

Table 3.11.2

| | $T'_0$ | | | | | |
|---|---|---|---|---|---|---|
| Sample size | 0 | 0.5 | 1.0 | 1.5 | 2.0 | 2.5 |
| $n = 20$ | | | | | | |
| 50 | | | | | | |
| | All accept $H_1$ (at $\alpha = 0.01$ level) | | | | | |
| 100 | | | | | | |
| 200 | | | | | | |

Thus there is little doubt that $\tilde{\theta}$ is better than $\hat{\theta}$.

*Example 3.11.4:* A person wishes to test the balance of a die. His intuitive decision rule is to claim that the die is not balanced if

$$\sum_{i=1}^{6} \left( o_i - \frac{n}{6} \right)^2 > c \quad ,$$

where $n$ is the number of tosses made and $o_i$ is the number of appearances of face $i$. He will toss the die 1200 times and the risk he can afford to make a false claim (that the die is unbalanced) is 0.05. Determine the value of $c$ for him.

*Solution:* From the problem we know that he is doing hypotheses testing on the two hypotheses:

$H_0$ : The die is balanced

$H_1$ : The die is not balanced.

We wish to find a value of $c$ such that

$$p(c) = \Pr\left\{ \sum_{i=1}^{6} \left( o_i - \frac{n}{6} \right)^2 > c \,|\, H_0 \right\} = 0.05 \quad .$$

Since the probability $p(c)$ is small, we need greater accuracy for $p(c)$ in our simulation. Let the simulation size $n_s$ be large enough to estimate $p(c)$ within an accuracy of $\pm 0.01$ with 95% confidence. Then the simulation size is (using 3.6.3 with $p \simeq 0.05$)

$$n_s = \frac{(1.96)^2 (0.05)(0.95)}{(0.01)^2} = 1825 \quad .$$

The simulation result is given in Table 3.11.3 and Fig. 3.11.1.

Table 3.11.3

| $c$: | 100 | 150 | 200 | 250 |
|---|---|---|---|---|
| $p(c)$: | 0.43 | 0.21 | 0.08 | 0.03 |

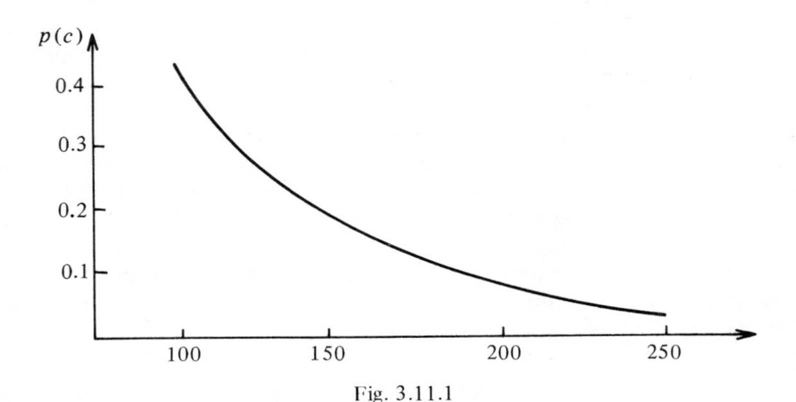


Fig. 3.11.1

By interpolation, we find $c \simeq 230$.

**Exercise 3**

3.1 A sample of 100 small stores was randomly taken to estimate the proportion of small stores that own computers. Twenty stores said yes. Estimate the true proportion by a 95% confidence interval.

3.2 A computer simulation will be used to assess the effectiveness of an automatic missile tracking system. The effectiveness will be evaluated by the probability of hitting the target. If we wish to estimate this probability within a range of 0.05 with 99% confidence, determine the sample size.

3.3 A random check of 20 electronic components from a huge shipment revealed that one of them is defective. Can one conclude that less than 10% of the whole shipment is defective?

3.4 A machine in a factory must be repaired if it produces more than 5% defectives. A random sample of 300 items from the product line contained 25 defective items. Does the sample strongly (define) suggest that the machine needs to be repaired?

3.5 A coin was tossed 20 times and 6 heads and 14 tails were observed. Can we say that this coin is not balanced? Answer this question based on the following two conditions prior to our experiment.
(a) We already suspect that this coin produces less heads than tails, and
(b) we just want to check the balance of this coin. (The risk we can take is less than 1 mistake for 10 decisions.)

3.6 A random digit generator is supposed to generate the 10 digits with the same probability. A 30-digit number was generated and there was no 0 in it. Can we say that this generator is biased against 0? Answer the question under two conditions. Before we look at the number,
(a) we already suspect that this generator may generate too few 0's, and
(b) we just wish to test the biasness of this generator.

3.7 In the past a certain kind of computer chips was known to have a yield rate of 10%. After implementing some new idea to improve the output quality, the manufacturer wished to determine whether the new method is effective by taking a sample. The manager decided that the probability of claiming improvement when there is none should be no more 0.05. On the other hand, if the new method can increase the yield from 10% to 15%, he wishes that the sample should have at least 0.95 probability to discover it. Determine the sample size for him.

3.8 Let $X$ be the number of defectives in a sample of $n$ items from a large population in which each item has the probability $p$ of being defective.

Find $\Pr\{X = x\}$ for

(i) $n = 12$, $p = 0.9$, $x = 10$;

(ii) $n = 10{,}000$, $p = 0.0002$, $x = 2$; and

(iii) $n = 500$, $p = 0.48$, $x = 260$.

(Hint: Use the right approximation formula.)

3.9 The acceptable defect rate of transistors is 2%. A large shipment has been received and the quality control engineer decides to take a random sample of 100 items, and accept this shipment if there are 2 or less defectives and reject it if there are three or more defectives.

(a) What is his risk of accepting an unacceptable shipment?

(b) What is his risk of rejecting a good lot? (A good lot is defined as one with defective rate $\leqslant 0.01$.)

(c) If he wants (a) to be $\leqslant 0.05$ and (b) to be $\leqslant 0.10$, should he change his sampling method? How?

Chapter 4

# COMPARISON OF STATISTICAL PROCEDURES

## 4.1. Introduction

Suppose you want to know the number of registered cars in a small town. One way to estimate this number is to go to a parking lot and keep a record of the tag numbers of the cars you find with the town's alias, say, ABC. You may see the numbers as follows:

$$\text{ABC-0521, ABC-4210, ABC-1441, ABC-0971, ABC-1152} \qquad (4.1.1)$$

Can we use these numbers to make an estimate of the number of cars in this town? To do this, let us assume that the serial number ranges from 0001 to $\theta$, which is the number of cars registered in this town. Thus, our goal is to estimate $\theta$ based on the data set (4.1.1).

You may feel that this assumption is unrealistic because some of the earlier registered cars may have left this town. This caution is obviously valid but difficult to verify. Let us just assume that all the cars are still in town for simplicity. This is not an unreasonable assumption if all the tags were issued quite recently. Actually this type of analysis was used by the American Intelligence Agency during World War II to estimate the production of German military equipment from the serial numbers of their destroyed planes, tanks, guns, etc. When the war was over, it was found that their estimates were in general better than those from other intelligence sources (see Ref. 9).

Let us go back to our original problem in a simplified form. Suppose we have $\theta$ cars tagged as $1, 2, \ldots, \theta$ and we have randomly observed a sequence of $n$ cars with tag numbers $X_1, X_2, \ldots, X_n$. One intuitive way to estimate $\theta$ is to argue as follows. Since the average tag number is

$$(1 + 2 + 3 + \ldots \theta)/\theta = (\theta + 1)/2 \quad ,$$

then if we equate this number with the sample mean $\overline{X} = (X_1 + \ldots + X_n)/n$, we would have a good estimate of $\theta$, i.e.

$$(\hat{\theta}_1 + 1)/2 = \overline{X} \quad \text{implies} \quad \hat{\theta}_1 = 2\overline{X} - 1 \ . \tag{4.1.2}$$

However, is (4.1.2) the *best* estimate of $\hat{\theta}$ based on the information we have? As in Example 3.11.3, can someone find a better estimator? Once this question arises, we will not feel comfortable if we cannot defend our estimator and claim that it is the best. In this chapter, we will study the criteria that determine the best estimators.

Before we discuss the statistical criteria, we will introduce a very basic criterion for a good estimator: *invariance*. An estimator (sometimes called a statistic) is said to be invariant under a transformation if its statistical inference will not be affected by presenting the data under the transformed values. For example, let $\overline{X}$, $S^2$, $n$ denote the sample mean, sample variance, and sample size, respectively. Recall that $T = \sqrt{n}\,(\overline{X} - \mu)/S$ in (3.11.5) has certain nice properties in doing simulation. We say that $T$ is invariant under *scale transformations* if our conclusion is independent of whether the heights are measured in inches or in centimetres. In this same example, if $\mu$ denotes the mean body temperature of that population, then we wish $T$ to be invariant whether the temperatures are measured in centigrade, Kelvin, or Fahrenheit. Since these temperatures differ by scale and starting points (0°), we actually wish $T$ to be invariant under *scale and location transformations*. Most tests we have discussed are invariant under various reasonable representations of the data. But we should always be careful, because sometimes a reasonable statistic may not be invariant (see Ex. 4.5 and 4.7).

## 4.2. The Minimum Variance Unbiased Estimator

To claim that an estimator $\hat{\theta}$ for a parameter $\theta$ is the best, we have to state clearly what we mean by 'the best'. The first commonly used criterion for a good estimator is *unbiasedness*. In other words, we require the expected value of $\hat{\theta}$ to be $\theta$. Or, following the notation of § 2.5, we require $E\hat{\theta} = \theta$. The reason for this requirement is obvious, because if we wish to estimate $\theta$, we certainly do not want the average value of our estimator $\bar{\theta}$ to be $2\theta$ or $\theta + 1$. Hence, if we have two estimators, $\hat{\theta}_1$ and $\hat{\theta}_2$, for $\theta$ based on the same sample, and one of them is unbiased while the other one is biased, we usually take the unbiased one.

Now suppose there are two unbiased estimators $\hat{\theta}_1$ and $\hat{\theta}_2$ for $\theta$, then what shall we do? Intuitively, we should choose the one with the smaller variance, i.e. if

$$E\hat{\theta}_1 = E\hat{\theta}_2 = \theta, \quad \text{and} \quad V(\hat{\theta}_1) < V(\hat{\theta}_2) \quad ,$$

then we choose $\hat{\theta}_1$. The reason for this choice is again obvious, because by the definition of variance $V(\hat{\theta}) = E(\hat{\theta} - \theta)^2$, we see that $\hat{\theta}_1$ is expected to be closer to $\theta$.

To extend our argument further, we see that we would have reached our ultimate goal if we can find an estimator $\hat{\theta}$ of $\theta$ which has the minimum variance among all the unbiased estimators. Our search usually stops when a *minimum variance unbiased estimator* (*MVUE*) is found. You may ask what's next if two minimum variance unbiased estimators are found. Should we check their third moments? Fortunately, almost all MVUE's are unique. But unfortunately, *to prove* that an estimate is MVUE is beyond the scope of this book and it will not be pursued further.

*Example 4.2.1:* Serial number analysis. It can be shown that the MVUE of $\theta$ is not (4.1.2) but

$$\hat{\theta}_2 = \{(n+1)X_{(n)}/n\} - 1 \quad , \tag{4.2.1}$$

where $X_{(n)} = \max(X_1, X_2, \ldots, X_n)$. It can shown further that both the estimate $\hat{\theta}_1 = 2\bar{X} - 1$ defined in (4.1.2) and $\hat{\theta}_2$ defined in (4.2.1) are unbiased and their variances are:

$$\text{Var}(\hat{\theta}_1) = \frac{(\theta+1)(\theta-n)}{3n} \quad ,$$

$$\text{Var}(\hat{\theta}_2) = \frac{(\theta+1)(\theta-n)}{n(n+2)} \quad . \tag{4.2.2}$$

It is quite striking that for large $n$, $\text{Var}(\hat{\theta}_2)$ is much smaller than $\text{Var}(\hat{\theta}_1)$, and consequently, $\hat{\theta}_1$ is an inefficient estimator.

It can be shown that the previously used estimators, $\bar{X}$ for $\mu$, $S^2$ for $\sigma^2$ are MVUE's under the normality assumption and the sample proportion $\hat{p}$ for the population proportion $p$ is also the MVUE under the binomial assumption.

## 4.3. The Mean Square Error and Efficiency

When simulation is used to compare estimators, the bias of an estimator cannot be ascertained. We may be able to demonstrate that the bias is small, but we cannot prove the bias is *absolutely 0* by simulation. Moreover, people usually prefer an estimator with a small bias and small variance to the one with 0 bias but a large variance. A combined criterion for bias and variance is called the *mean square error* which is defined by

$$\mathrm{MSE}(\hat{\theta}) = E(\hat{\theta}-\theta)^2 \quad , \tag{4.3.1}$$

where $\hat{\theta}$ is an estimate of $\theta$. It can be shown that

$$\mathrm{MSE}(\hat{\theta}) = \mathrm{Var}(\hat{\theta}) + (E\hat{\theta}-\theta)^2 \quad .$$

Thus for two unbiased estimators, comparing MSE is the same as comparing their variances.

To compare two estimators based on their MSEs is easy by simulation. One need only to simulate a large number of $\hat{\theta}$ and evaluate the MSE by the average of $(\hat{\theta}-\theta)^2$. In Table 4.3.1 are some comparisons of $\hat{\theta}_1$ and $\hat{\theta}_2$ defined by (4.1.2) and (4.2.1) based on $\theta = 4261$ and 1,000 simulations. The superiority of $\hat{\theta}_2$ is obvious.

Table 4.3.1

| $n =$ | $\mathrm{MSE}(\hat{\theta}_1)/10^4$ | $\mathrm{MSE}(\hat{\theta}_2)/10^4$ |
|---|---|---|
| 10 | 62.3 | 15.2 |
| 20 | 31.4 | 4.2 |
| 50 | 13.6 | 0.7 |

The *efficiency of one estimator* $\hat{\theta}_1$ *relative to* $\hat{\theta}_2$ is defined by

$$\mathrm{eff}(\hat{\theta}_1/\hat{\theta}_2) = \frac{1/\mathrm{MSE}(\hat{\theta}_1)}{1/\mathrm{MSE}(\hat{\theta}_2)} = \frac{\mathrm{MSE}(\hat{\theta}_2)}{\mathrm{MSE}(\hat{\theta}_1)} \quad . \tag{4.3.2}$$

We often label $\hat{\theta}_1$ and $\hat{\theta}_2$ so that $\hat{\theta}_2$ is the one with smaller MSE. Thus $\mathrm{eff}(\hat{\theta}_1/\hat{\theta}_2)$ is usually a number between 0 and 1. In Table 4.3.1 we see that the efficiency of $\hat{\theta}_1$ relative to $\hat{\theta}_2$ at $n = 50$ is $0.7/13.6 = 0.0514 = 5.14\%$.

You may wonder why we need efficiency if we always choose the estimator with the smaller (the smallest) MSE. The estimators with larger MSE will not be used anyway. But this is not always the case. Sometimes an estimator with larger MSE has some other nice properties (mainly robustness, see § 4.8) that the estimator with smaller MSE does not have. People sometimes would rather choose a highly efficient robust estimator than a non-robust MVUE.

There are also situations where the MVUE cannot be found but the minimum variance itself can be evaluated. Thus, we can use the efficiency criterion to determine how close an estimator is to MVUE.

*Example 4.3.1:* In a normal population, $N(\mu, \sigma^2)$, we know that $\overline{X}$ is the MVUE of $\mu$ with variance $V(\overline{X}) = \sigma^2/n$. Suppose the *sample median* $\hat{m}$ is used as an estimator for $\mu$. What is its efficiency? The *sample median* $\hat{m}$ is defined as the center value of the ordered sample if $n$ is odd, and the average of the two center values if $n$ is even.

It can be easily seen that the efficiency is invariant under scalar and location transformation (see Ex. 4.6). Thus we may draw a sample from $N(0, 1)$. The following results are found by 1,000 simulations (Fig. 4.3.1).

Note: $\text{eff}(\hat{m}) \equiv \text{eff}(\hat{m}/\overline{X}) = \text{MSE}(\overline{X})/\text{MSE}(\hat{m}) = 1/[n\text{MSE}(\hat{m})]$.

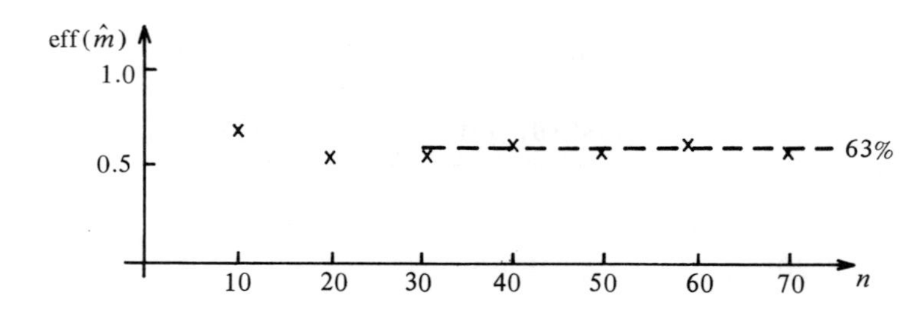


Fig. 4.3.1. eff($\hat{m}$) in a normal population, $n$ = sample size.

*Example 4.3.2:* Suppose we have a sample from a Cauchy distribution with the density function

$$f(x) = \frac{1}{\pi(1 + (x - \theta)^2)}, \qquad -\infty < x < \infty .$$

We wish to estimate $\theta$ from a sample, $X_1, X_2, \ldots, X_n$. It can be shown by statistical theory that the minimum possible variance for $\hat{\theta}$ is $2/n$ (Johnson & Kotz [13], p. 158), but to find $\hat{\theta}$ is very difficult. Since $\theta$ is at the center of the symmetry density function $f(x)$, we may try the sample mean and the sample median. Figure 4.3.2 gives some simulation results. The rest of the curves are left as an exercise.

We see that the mean is a very inefficient estimate of $\theta$ but the median is quite acceptable.

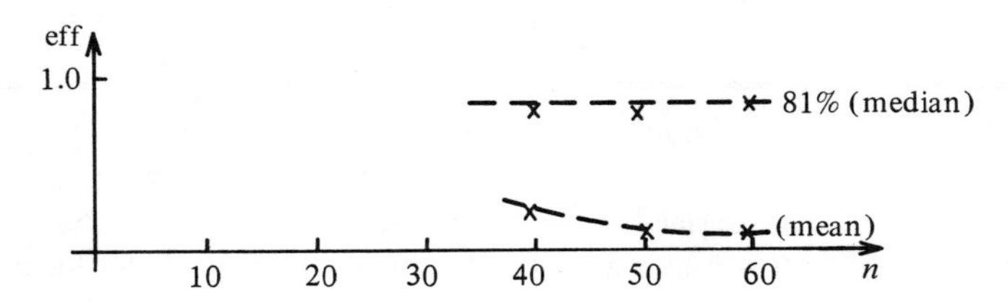


Fig. 4.3.2. The efficiencies of the mean and median (relative to the minimum possible MSE) for a Cauchy distribution.

## 4.4. The Maximum Concentration Criterion

An even stronger criterion than MSE is the concentration criterion. Let $\hat{\theta}_1$ and $\hat{\theta}_2$ be two estimators of $\theta$. We say that $\hat{\theta}_1$ has a higher concentration than $\hat{\theta}_2$ if for all $\epsilon > 0$

$$\Pr\{|\hat{\theta}_1 - \theta| < \epsilon\} \geqslant \Pr\{|\hat{\theta}_2 - \theta| < \epsilon\} \quad . \tag{4.4.1}$$

It can be shown that higher concentration implies smaller MSE but not vice versa. What (4.4.1) states is that $|\hat{\theta}_1 - \theta| < \epsilon$ is more likely to occur than $|\hat{\theta}_2 - \theta| < \epsilon$. In the MSE criterion, $\text{MSE}(\hat{\theta}_1) < \text{MSE}(\hat{\theta}_2)$ may imply $\Pr\{|\hat{\theta}_1 - \theta| < \epsilon\} \geqslant \Pr\{|\hat{\theta}_2 - \theta| < \epsilon\}$ for some $\epsilon$ and $\Pr\{|\hat{\theta}_1 - \theta| < \epsilon\} \leqslant \Pr\{|\hat{\theta}_1 - \theta| < \epsilon\}$ for some other $\epsilon$.

A weaker criterion on concentration is to evaluate

$$p = \Pr\{|\hat{\theta}_1 - \theta| \leqslant |\hat{\theta}_2 - \theta|\} \tag{4.4.2}$$

and claim that $\hat{\theta}_1$ is better if $p > 0.5$. Though (4.4.1) and (4.4.2) have not received too much attention in statistical literature due to theoretical difficulties, they are quite easy to evaluate by simulations.

## 4.5. Other Statistical Inference Procedures

Let us summarize the efficiencies of the mean and median in the 3 examples we have discussed in the previous sections (see Table 4.5.1).

Here we see that if the probability structure of a population is unknown, it is quite difficult to determine the best possible estimator for a given parameter. Unfortunately, this is the case in many practical situations, especially when the survey is done for the first time. Because of this concern, statistical inference

Table 4.5.1. Efficiencies of the mean and median in three distributions. (The values are theoretical results for large $n$.)

| | Population & Parameter | | |
|---|---|---|---|
| Estimator based on | Serial # $\theta$ = pop. size | Normal $\theta$ = mean | Cauchy $\theta$ = center location |
| mean | $3/(n+2)$ | 1 | 0 |
| median | $1/n$ | $2/\pi$ | 0.81 |

has been divided into the following four main categories.

1) Parametric inference: The detailed probability structure of the population is assumed known, except a few parameters which must be estimated. For example, we may assume the population is normal with unknown mean $\mu$ and unknown variance $\sigma^2$.
2) Nonparametric inference: The probability structure is assumed to be unknown, except a few trivial conditions such as continuity.
3) Robust inference: The probability structure is assumed to be known to a great extent, but a safeguard of possible exceptions is considered. For example, we may assume the population is normal, but we consider it possible to have keypunch errors in data processing.
4) Density estimation: Statistical inference is made by estimating the density function of the population.

In this book, we will discuss the first three methods by examples.

## 4.6. Parametric Statistical Inference

The most commonly used method in parametric inference is maximum likelihood estimation. The basic idea is to find the most likely parameter(s) that makes the data $X_1, X_2, \ldots, X_n$ have the best chance of occurring. We define the likelihood function for a parameter(s) as the product of the probability or the probability density function for the given data. The maximum likelihood estimate (m.l.e.) of the parameter is the value of the parameter that maximizes the likelihood function. For example, let $X_1, X_2, \ldots, X_n$ denote the outcomes of $n$ tosses of a coin. Let $X_i = 1$ if the $i$th toss is a head and $X_i = 0$ if the $i$th toss is a tail. Then the probability of $X_i$ being $x_i$ is $p^{x_i}q^{1-x_i}$, where $p$ is the probability of getting a head and $q = 1 - p$, and $X_i$ is the random variable associated with the coin and $x_i$ is the datum. Thus, by definition, the likelihood of this experiment is

$$\begin{aligned} L(p \mid x_1, \ldots, x_n) &= p^{x_1} q^{1-x_1} \cdot p^{x_2} q^{1-x_2} \ldots p^{x_n} q^{1-x_n} \\ &= p^{\Sigma x_i} q^{n-x_i} \\ &= p^{\Sigma x_i} (1-p)^{n-\Sigma x_i} \quad . \qquad (4.6.1) \end{aligned}$$

To maximize (4.6.1) is the same as maximizing $\log_e L = \ln L$. Since

$$\ln L = \Sigma x_i \ln p + (n - \Sigma x_i) \ln (1-p) \quad , \qquad (4.6.2)$$

by differentiation, we see that $\ln L$ is maximized at $\hat{p} = \Sigma x_i / n$, which is the sample proportion.

Take the exponential distribution as another example. Suppose $x_1, x_2, \ldots,$ $x_n$ come from an exponential distribution with the density function

$$f(x) = \lambda e^{-\lambda x} \quad , \qquad 0 < x < \infty \quad .$$

Then according to the definition of the likelihood function

$$\begin{aligned} L(\lambda \mid x_1, x_2, \ldots, x_n) &= \lambda e^{-\lambda x_1} \cdot \lambda e^{-\lambda x_2} \ldots \lambda e^{-\lambda x_n} \\ &= \lambda^n e^{-\lambda \Sigma x_i} \quad . \qquad (4.6.3) \end{aligned}$$

By differentiating $\ln L$ with respect to $\lambda$, we may find that (4.6.3) is maximized at

$$\hat{\lambda} = (\bar{X})^{-1} \quad .$$

Thus $(\bar{X})^{-1}$ is the m.l.e. of $\lambda$.

It can be shown that $\bar{X}$ is the m.l.e. of $\mu$ in a normal sample. The m.l.e. of $\sigma^2$ in a normal sample $X_1, X_2, \ldots, X_n$ when the mean is unknown is not the commonly used $S^2$, but

$$S'^2 = \frac{\sum_{i=1}^{n} (X_i - \bar{X})^2}{n} = \frac{n-1}{n} S^2 \quad .$$

There is still controversy as to which of the two estimators $S^2$ or $S'^2$ should be used for $\sigma^2$ in a normal sample. While $S^2$ is unbiased, $S'^2$ has a smaller mean square error. Fortunately, for large $n$, the difference is negligible.

## 4.7. Non-parametric Statistical Inference

As we have discussed in §4.5 a non-parametric procedure does not assume any specific form of the population distribution. We may still wish to make inference on some parameters, but these parameters no longer appear in simple forms in a density function. Let us use the commonly heard median income as an example. Incomes usually follow a very skewed distribution such as Fig. 4.7.1.

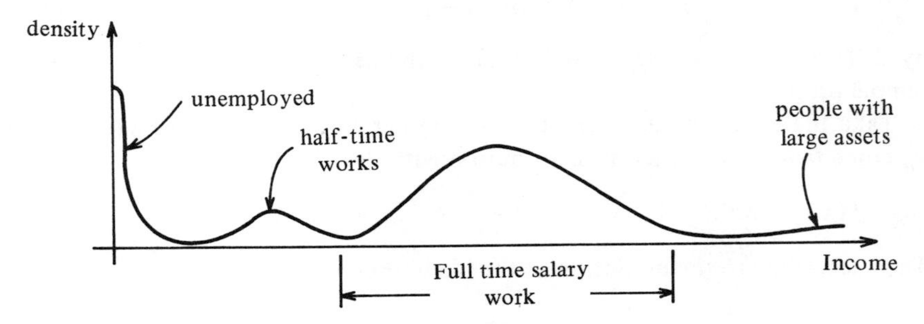


Fig. 4.7.1 A typical income distribution.

This distribution is obviously difficult to parametrize. Moreover, the mean income of a group can sometimes be very misleading. For example, a block has 10 houses with 9 salaried workers with annual income around 20K each and one rich person with income of 1 million per year. Thus the mean income, which is 118K, of this block obviously tells us very little about the affluence. However, if we state that the block has a median income of 17,000 per year, it gives us a better picture, because we know that half of the people have income more than 17,000 and the other half have income below this value. That is why people usually use median income as an indicator of the wealth of a community. A natural way to estimate the median is to use the sample median $\hat{m}$ which is defined in §4.3. We see that little has been assumed about the distribution form of the population.

To find a confidence interval for $m$, we need to order the data $x_1, x_2, \ldots x_n$ as

$$x_{(1)} < x_{(2)} < \ldots < x_{(n-1)} < x_{(n)} \quad , \tag{4.7.1}$$

where $x_{(1)}, x_{(2)}, \ldots, x_{(n)}$ is the ascending arrangement of $x_1, x_2, \ldots, x_n$. Then it can be shown that for a large population

$$\Pr\{X_{(r)} \leqslant m \leqslant X_{(s)}\} = \sum_{i=r}^{s-1} \binom{n}{i} \left(\frac{1}{2}\right)^n \tag{4.7.2}$$

Thus, a $(1-\alpha)$ confidence interval for $m$ is formed by choosing $r$ and $s$ in (4.7.2) such that $(s-r)$ has the smallest value subject to

$$\sum_{i=r}^{s-1} \binom{n}{i} \left(\frac{1}{2}\right)^n \geqslant 1-\alpha \quad . \tag{4.7.3}$$

Since the binomial distribution $\text{Bi}(n, 0.5)$ is symmetric with respect to $0.5n$, the $s$ in (4.7.2) and (4.7.3) should be chosen as $s = n - r + 1$. Thus (4.7.3) implies that we use the largest $r$ such that

$$\sum_{i=r}^{n-r} \binom{n}{i} \left(\frac{1}{2}\right)^n \geqslant 1-\alpha \quad ,$$

or the largest $r$ such that

$$2\sum_{i=0}^{r-1} \binom{n}{i} \left(\frac{1}{2}\right)^n \leqslant \alpha \quad . \tag{4.7.4}$$

*Example 4.7.1:* Let $n = 10$. Find a 95% confidence interval for the median.

*Solution:* In this case $\alpha = 0.05$, $n = 10$. It is very easy to see that for

$$r = 1 \;, \qquad 2\sum_{i=0}^{0} \binom{n}{i} \left(\frac{1}{2}\right)^n = 2\left(\frac{1}{2}\right)^{10} = 0.002$$

$$r = 2 \;, \qquad 2\sum_{i=0}^{1} \binom{n}{i} \left(\frac{1}{2}\right)^n = 2(1+10)\left(\frac{1}{2}\right)^{10} = 0.022$$

$$r = 3 \;, \qquad 2\sum_{i=0}^{2} \binom{n}{i} \left(\frac{1}{2}\right)^n = 2(1+10+45)\left(\frac{1}{2}\right)^{10} = 0.109.$$

Thus a 95% confidence interval for $m$ is $(X_{(2)}, X_{(9)})$. Actually, this is also a 97.8% confidence interval.

*Example 4.7.2:* Solve the same problem for $n = 25$ in the previous example.

*Solution:* Since large combinatoric numbers are involved, we should do it using a computer. It can be shown that for

$$r = 8 \quad , \qquad 2\sum_{i=0}^{7}\binom{n}{i}\left(\frac{1}{2}\right)^n = 0.044,$$

and for

$$r = 9 \quad , \qquad 2\sum_{i=0}^{8}\binom{n}{i}\left(\frac{1}{2}\right)^n = 0.108 \quad .$$

Thus a 95% confidence interval for $m$ when $n = 25$ is $(X_{(8)}, X_{(18)})$ by taking $r = 8$.

Actually, it can be shown that for large $n$,

$$r = \frac{n+1}{2} - \frac{\sqrt{n}}{2} z_{1-\alpha/2} \tag{4.7.5}$$

for a $(1-\alpha)$ confidence interval for $m$. Take $n = 25$ for example, the $r$ for $\alpha = 0.05$ is $r = (26)/2 - (\sqrt{25}/2)1.96 = 8.1 \approx 8$.

In addition to the median, we define $\xi_p$ as the population quantile of order $p$ by

$$p = \Pr\{X \leqslant \xi_p\} \quad ,$$

where $X$ is a random variable from the population. The point estimate for $\xi_p$ is the lower $p100\%$ sample quantile. A $(1-\alpha)$ confidence interval for $\xi_p$ is $(X_{(r)}, X_{(s)})$ where $r$, $s$ minimize $|s-r|$ subject to

$$\sum_{i=r}^{s-1}\binom{n}{i} p^i(1-p)^{n-i} \geqslant 1-\alpha \quad . \tag{4.7.6}$$

(see David [4], pp. 15-16).

## 4.8. Robust Statistical Procedures

Robust statistical procedures are estimators or tests that are insensitive to small deviations from the assumptions. For example, suppose we have data from a normal population $N(\mu, \sigma^2)$. While we are pretty sure the population is normal, we are not so confident that there were no errors made during data collection and recording. Suppose the data are, for simplicity,

$$1, 2, 3, 4, 5, 6, 7, 8, 9, 10 \ . \tag{4.8.1}$$

We know that the sample mean 5.5 is the MVUE for $\mu$. Now suppose there is a keypunch error that 9 was keypunched as 90. Then the sample mean 13.6 will be a very bad estimate of $\mu$. What should we do in this situation? For a small sample size such as this, we may suspect that the extremely large value 90 was due to some kind of error and it may demand some investigation. But for a large data set, especially when the data are to be analyzed by a computer, a standard procedure is necessary. This procedure should be able to reduce the effect of 'bad' values, yet not lose too much efficiency when the data is good. One of these procedures is called the trimmed mean.

A $\delta\%$ trimmed mean is the mean obtained after discarding the upper $\delta\%$ of the data and the lower $\delta\%$ of the data. For example, a 10% trimmed mean for the data in (4.8.1) is

$$\hat{m}(10\%) = \frac{1}{8}(2+3+4+5+6+7+8+9) = 5.5 \ ,$$

and for the miskeypunched data

$$\hat{m}(10\%) = \frac{1}{8}(2+3+4+5+6+7+8+10) = 5.625 \ .$$

They are pretty close. Note here we are trading efficiency for security. Discarding data is equivalent to reducing the sample size, which consequently reduces the efficiency of our estimator. But it protects the estimate from extreme values. For large $n$, this protection out-weighs the loss of efficiency. For example, when $n = 100$, a 5% trimmed mean should give us sufficient protection while the loss of efficiency is negligible.

When the trimmed mean is used, the $(1-\alpha)$ confidence interval becomes

$$\hat{m}_{(\delta\%)} \pm t_{\nu,\,1-\alpha/2} \sqrt{\frac{SS_\delta}{m(m-1)}}$$

where $\nu = n - 2m - 1$, $\hat{m}_{(\delta\%)}$ is the trimmed mean and $SS(\delta\%)$ is called the

Winsorized sum of squares defined by

$$SS_\delta = m(X_{(m+1)} - \hat{m}(\delta\%))^2$$

$$+ \sum_{i=m+1}^{n-m} (X_{(i)} - \hat{m}(\delta\%))^2 + m(X_{(n-m)} - m(\delta\%))^2 \quad ,$$

where $x_{(1)} < x_{(2)} < x_{(3)} < \ldots < x_{(n)}$ represents the ordered data. (See reference [1], p. 116).

**Exercise 4**

4.1 Let the population be 1, 2, . . . , $N$ and $X_{(n)}$ be the maximum of a sample of size $n$. Show that

(i) $\Pr\{X_{(n)} = i\} = \binom{i-1}{n-1} \Big/ \binom{N}{n}$, $i = n, n+1, \ldots, N$.

(ii) Use (i) to show that $\hat{\theta}_2$ defined in (4.2.1) is unbiased.

4.2 Use the normal approximation to the binomial distribution to show (4.7.5). (Hint: From (4.7.4) and $\Pr\{Z \leqslant (r - n/2 - 1/2)/\sqrt{(n/4)}\} = \alpha/2$.)

4.3 (*Programming exercise*) Use the inverse CDF method to generate Cauchy random variates defined in Example 4.3.2 and complete Fig. 4.3.2.

4.4 Find the maximum likelihood estimates of $\mu$ and $\sigma^2$ from a sample $x_1$, $x_2$, . . . $x_n$ from a normal population.

4.5 A total of $n$ homing pigeons will be released at a location far away from their home. When they are released, the direction of their flight will be recorded as $\alpha_1, \alpha_2, \ldots, \alpha_n$, in degrees by letting north be 0 degrees. All the angles are measured between 0 and 360 degrees clockwise. Suppose we make statistical inference on their flight direction by the arithmetic mean $\bar{\alpha} = \Sigma\alpha_i/n$.

(i) Is this quantity $\bar{\alpha}$ invariant under the shift of the starting angle $0°$?

(ii) What is the practical significance of this result?

4.6 Suppose the median $\hat{m}$ is used as an alternative to estimate the mean $\mu$ from a normal population. To compare it with the sample mean $\bar{X}$, the concentration criterion will be used, i.e. we intend to estimate

$$p = \Pr\{|\hat{m} - \mu| < |\bar{X} - \mu|\}$$

by computer simulation. Can we consider the population as $X \sim N(0, 1)$ without loss of generality?

4.7 Suppose $X$ is a random variable uniformly distributed in $[a, b]$. Then the range of $X$ is defined as $R \equiv b - a$. Suppose $X_1, X_2, \ldots, X_n$ are observed and the range $R$ will be estimated by $\hat{R} = \max(X_1, \ldots, X_n) - \min(X_1, \ldots, X_n)$. We wish to find

$$p = \Pr\{|\hat{R} - R| < \epsilon\}$$

by simulation. Can we let $[a, b] = [0, 1]$ without loss of generality?

4.8 The number of breakdowns of a computer in a week follows a Poisson distribution $P_0(\lambda)$. Let $X_1, X_2, \ldots, X_n$ denote the number of breakdowns in weeks 1, 2, ..., $n$. Naturally we will estimate $\lambda$ by

$$\hat{\lambda} = \Sigma X_i / n \quad .$$

To assess the reliability of $\hat{\lambda}$, simulation will be used to estimate

$$p = \Pr\{|\hat{\lambda} - \lambda| < \epsilon\} \quad .$$

Thus, a table for $p$ will contain three parameters, $n$, $\epsilon$, and $\lambda$. Can the number of parameters be reduced?

Chapter 5

# ELEMENTARY STATISTICAL PROCEDURES

## 5.1. General Formulation

In this chapter, a summary of the procedures discussed in the previous chapters and some additional commonly used elementary statistical procedures are given. These procedures were developed before the wide availability of computers. Hence their statistical properties were derived not by simulation but by analytic methods. Most of the populations are restricted to simple distributional forms such as binomial or normal, but with the help of a computer, their applicability to other populations can be easily evaluated by simulation.

The basic idea behind all these statistical procedures is quite simple. Suppose we wish to make a statistical inference about a parameter $\theta$. No matter whether the statistical inference involves hypotheses testing or confidence intervals, the first task is to find a reasonable estimate $\hat{\theta}$ for $\theta$. If we are fortunate enough to find the distribution of $\hat{\theta}$, or the distribution of a function of $\hat{\theta}$, say $f(\hat{\theta})$, or even an approximation to them, then the rest of the statistical inference will be routine. For example, suppose we find the distribution for $W = \hat{\theta} - \theta$. Then for any given $1-\alpha$, we can find an $\epsilon$ such that $\Pr\{|W| < \epsilon\} = 1-\alpha$, or equivalently,

$$\Pr\ |\hat{\theta} - \theta| \leqslant \epsilon \ = 1-\alpha \tag{5.1.1}$$

This means that a $(1-\alpha)$ confidence interval for $\theta$ is $\hat{\theta} \pm \epsilon$. On the other hand, if we wish to test

$$\begin{array}{llll} \text{(i)} \ H_0 : \theta = \theta_0 \quad (\theta \leqslant \theta_0) & \text{vs} & H_1 : \theta > \theta_0, & \text{or} \\ \text{(ii)} \ H_0 : \theta = \theta_0 \quad (\theta \geqslant \theta_0) & \text{vs} & H_1 : \theta < \theta_0, & \text{or} \\ \text{(iii)} \ H_0 : \theta = \theta_0 & \text{vs} & H_1 : \theta \neq \theta_0 & \end{array} \tag{5.1.2}$$

for a given $\theta_0$, then our corresponding decision rules for accepting $H_1$ in the above three hypotheses are respectively (i) $\hat{\theta} > c_1$, (ii) $\hat{\theta} < c_2$, and (iii) $|\hat{\theta} - \theta_0| > c_3$ for some $c_1, c_2, c_3$.

The relationships between the risk $\alpha$ and $c_1$, $c_2$, and $c_3$, respectively, are given by

$$\text{(i)} \quad \alpha = \Pr\{\hat{\theta} < c_1 | H_0\} = \Pr\{\hat{\theta} - \theta_0 < c_1 - \theta_0\}$$
$$= \Pr\{W < c_1 - \theta_0\} \quad ,$$

$$\text{(ii)} \quad \alpha = \Pr\{W > c_2 - \theta_0\} \text{ , and}$$

$$\text{(iii)} \quad \alpha = \Pr\{|W| > c_3\} \quad . \qquad (5.1.3)$$

Since $W$ has a known distribution, we can find $\alpha$ for a given value of $c$ and $c$ for a given $\alpha$. Moreover, the $p$-values, defined as the minimum risk such that $H_1$ is accepted at the present situation, are respectively

$$\text{(i)} \quad p\text{-value} = \Pr\{W > \hat{\theta}_e - \theta_0\}$$

$$\text{(ii)} \quad p\text{-value} = \Pr\{W < \hat{\theta}_e - \theta_0\}$$

$$\text{(iii)} \quad p\text{-value} = \Pr\{|W| > |\hat{\theta}_e - \theta_0|\}. \quad , \qquad (5.1.4)$$

where $\hat{\theta}_e$ denotes the estimate $\hat{\theta}$ from the given *data,* not the general $\hat{\theta}$ from *random variables*. The $p$-values can now be computed, since $\hat{\theta}_e$ and $\theta_0$ are known and $W$ has a known distribution.

*Remark 1.* The distribution of $\hat{\theta}$ or $f(\hat{\theta})$ usually depends on the true value $\theta$ and the sample size $n$. Let us denote this distribution by $W(\theta)$. There will be a problem in working with $W(\theta)$ and constructing a confidence interval, because $\theta$ is unknown. Sometimes we may substitute $\hat{\theta}$ for $\theta$ (i.e. use $W(\hat{\theta})$), but usually we cannot do this for a small sample size $n$. However, there is no problem for hypothesis testing because in (5.1.3) and (5.1.4), the distribution of $W$ is under the null hypothesis that $\theta$ has the given value $\theta_0$. That is why the hypotheses testing are exact but the confidence intervals are not when we try to make an inference about a proportion for small sample sizes (see Section 3.7).

*Remark 2.* The confidence interval problems are usually restricted to one or two unknown parameters. When the number of parameters is more than two, hypotheses testing is easier to apply. One example is the frequency table inference in Example 3.11.4. It is quite difficult to construct confidence intervals for six frequencies simultaneously, but hypothesis testing presents no similar problem.

It is not our intention to discuss the details of these tests with many examples in the following sections. There are many good books that cover these tests quite thoroughly. Among them are: Mendenhall, Scheaffer, and Wackerly [1980], McClave and Dietrich [1979], and Scheaffer and McClave [1982].

## 5.2. Inference on the Population Means

### *5.2.1 One sample*

a) Sample: $Y_1, Y_2, \ldots, Y_n$
   Assumption: The sample is from a normal population with unknown mean $\mu$ and unknown variance $\sigma^2$. In other words: $Y_1, Y_2, \ldots, Y_n \sim N(\mu, \sigma^2)$.

b) Parameter of interest: $\theta = \mu$.

c) A reasonable estimate for $\theta$: $\hat{\theta} = \overline{Y} = (Y_1 + Y_2 + \ldots + Y_n)/n$.

d) Known distribution of $\hat{\theta}$: $\sqrt{n}(\overline{Y} - \mu)/S$ has a $t$-distribution with $(n-1)$ degrees of freedom, where $S^2$ is the sample variance

$$S^2 = \sum_{i=1}^{n} \frac{(Y_i - \overline{Y})^2}{(n-1)}$$

$$= \frac{1}{n-1}\left\{\left(\sum_{1}^{n} Y_i^2\right) - n\overline{Y}^2\right\} . \qquad (5.2.1\text{d})$$

e) Remark: The distribution of this $t$-statistic $\sqrt{n}(\overline{Y} - \mu)/S$ is quite robust against non-normality.

### *5.2.2 Two sample*

a) Sample: $Y_{11}, Y_{12}, Y_{13}, \ldots, Y_{1n_1} \sim N(\mu_1, \sigma^2)$
   $Y_{21}, Y_{22}, Y_{23}, \ldots, Y_{2n_2} \sim N(\mu_2, \sigma^2)$ .
   Note: The variances are assumed to be the same.

b) Parameter of interest: $\theta = \mu_1 - \mu_2$.

c) A reasonable estimate for $\hat{\theta}$: $\theta = \overline{Y}_1 - \overline{Y}_2$, the difference of the two sample means.

d) Known distribution of $\hat{\theta}$: Let $S_1^2$ and $S_2^2$ be the sample variances from the first and the second samples respectively and the pooled variance

$$S^2 = \frac{(n_1 - 1)S_1^2 + (n_2 - 1)S_2^2}{n_1 + n_2 - 2} \quad ,$$

$$W = \frac{\hat{\theta} - \theta}{S(1/n_1 + 1/n_2)^{1/2}} = \frac{\bar{Y}_1 - \bar{Y}_2 - (\mu_1 - \mu_2)}{\sqrt{S^2(1/n_1 + 1/n_2)}} \quad , \tag{5.2.2d}$$

Then $W$ has a $t$-distribution with $(n_1 + n_2 - 2)$ degrees of freedom.

e) Remark: The distribution $W$ is also quite robust against non-normality and unequal variances if $n_1 \approx n_2$.

*Example 5.2.1:* The repair times of a large computer in 9 randomly picked repairs are recorded in hours as

32, 37, 35, 28, 41, 44, 35, 31, 34 .

Find a 95% confidence interval for the mean repair time.

*Solution:* Let us assume that it is reasonable to assume normality for the data. According to (5.2.1d) and (5.1.1), we have

$$\Pr\left\{\left|\frac{\sqrt{n}(\bar{Y} - \mu)}{S}\right| \leqslant t_{8,0.975}\right\} = 0.95 \quad .$$

From the $t$-table (Table A.2), we have $t_{8,0.975} = 2.365$. Thus a 95% confidence interval for $\mu$ is

$$|\bar{Y} - \mu| \leqslant 2.365 \cdot \frac{S}{\sqrt{n}} \quad ,$$

or

$$\bar{Y} \pm 2.365 \frac{S}{\sqrt{n}} \quad .$$

From the data we find that $\bar{y} = 35.22$, $s^2 = 24.445$, or $s = 4.944$. Thus, a 95% confidence interval for $\mu$ is $35.22 \pm 2.306 \times 4.944/\sqrt{9} = 35.22 \pm 3.90$ or (31.32, 39.12).

*Example 5.2.2:* Suppose the nine pieces of data in the previous example were the repair times by repairman A. There is also a sample from repairman B. The data, the sample mean, and variance are

| *Data* | $\bar{y}_2$ | $s_2^2$ | $n_2$ |
|---|---|---|---|
| 35, 31, 29, 25, 34, 40, 27, 32, 31, 32 | 31.6 | 17.82 | 10 |

Can we conclude with confidence that repairman B on the average takes less time to repair the computer than A?

*Solution:* Let $\mu_1$ and $\mu_2$ be the mean repair times for A and B. To set up the hypotheses, we have to know whether we only wish to test $\mu_2 < \mu_1$ or merely compare the two repairmen (in this case $H_1 : \mu_1 \neq \mu_2$) *before* the data were collected. Suppose before the data were collected, we wish to know which repairman is quicker but *after* looking at the data, we suspect that B is quicker. In this case, our hypotheses are

$$H_0 : \mu_1 = \mu_2 , \quad (\text{or } \theta = 0\text{: } \theta = \mu_1 - \mu_2)$$

$$H_1 : \mu_1 \neq \mu_2 , \quad (\text{or } \theta \neq 0) \quad .$$

The decision rule is to accept $H_1$ if $|\hat{\theta} - \theta_0| > c$ and the $p$-value is

$$\Pr\{|\hat{\theta} - \theta_0| > c \,|H_0\} \quad , \tag{5.2.3}$$

where $c$ is chosen so that $H_1$ is accepted at the lowest allowable risk, or $|\hat{\theta}_e - \theta_0|$ where $\hat{\theta}_e$ is defined in (5.1.4). From (5.2.2d), (5.2.3) becomes

$$\Pr\left\{\left|\frac{\bar{Y}_1 - \bar{Y}_2}{S\sqrt{(1/n_1 + 1/n_2)}}\right| > \frac{c}{S\sqrt{(1/n_1 + 1/n_2)}}\right\}$$

$$= \Pr\left\{|t_{n_1 + n_2 - 2}| > \frac{c}{S\sqrt{(1/n_1 + 1/n_2)}}\right\}$$

where

$$c = |\bar{y}_1 - \bar{y}_2| = |35.22 - 31.6| = 3.62$$

$$s^2 = \frac{(n_1 - 1)S_1^2 + (n_2 - 1)S_2^2}{n_1 + n_2 - 2} = \frac{8 \times 24.445 + 9 \times 17.82}{17}$$

$$= 20.94 \quad .$$

Thus the $p$-value for this test is

$$\Pr\left\{|t_{9+10-2}| > \frac{3.62}{\sqrt{20.94(1/9+1/10)}}\right\}$$

$$= \Pr\{|t_{17}| > 1.72\} \quad = 0.11$$

With the risk equal to 0.11, we usually cannot conclude which one of the two repairmen is better.

*5.2.3 Two sample, paired*

a) Sample: $\begin{pmatrix} Y_{11} \\ Y_{21} \end{pmatrix}, \begin{pmatrix} Y_{12} \\ Y_{22} \end{pmatrix}, \begin{pmatrix} Y_{13} \\ Y_{23} \end{pmatrix}, \ldots, \begin{pmatrix} Y_{1n} \\ Y_{2n} \end{pmatrix}$

Note: The data within each pair are obtained from a more homogeneous experimental condition than the data between pairs. The assumptions are $Y_{1i} \sim N(\mu_1, \sigma_1^2)$, $Y_{2i} \sim N(\mu_2, \sigma_2^2)$ and the data within a pair can be correlated.

b) Parameter of interest: $\theta = \mu_1 - \mu_2$.

c) A reasonable estimate for $\theta$: Let $D_i = Y_{1i} - Y_{2i}$, $i = 1, 2, \ldots, n$, $\hat{\theta} = \overline{D} = \overline{Y}_1 - \overline{Y}_2$.

d) Since $D_i$'s can be considered as one sample data, the distribution of $\hat{\theta}$ is the same as that of $\hat{\theta}$ in (5.2.1d) when the $Y$'s are replaced by the $D$'s.

*Example 5.2.3:* The breakdown time in five months of two computers are recorded as follows. Test the hypothesis that computer A breaks down more often than B. The reason that a one-sided test is used is because A is older and we expect a possible higher breakdown rate.

| Month | 1 | 2 | 3 | 4 | 5 |
|---|---|---|---|---|---|
| A | 30 | 34 | 26 | 33 | 28 (in hours) |
| B | 28 | 29 | 26 | 30 | 24 (in hours) |

*Solution:* Since the breakdowns of a computer depend on the room temperature, humidity, incoming voltage variation, etc., we probably should think that comparing the computers within a month is fairer than between months. Thus we consider the data being paired by month. The monthly differences are 2, 3, 0, 3, 4. Thus $\bar{d} = 2.8$, and $s_d^2 = 3.7$, $s_d = 1.92$, and the $p$-value of this test is

$$\Pr\left\{t_4 > \frac{2.8}{1.92/\sqrt{5}}\right\} = \Pr\{t_4 > 3.25\} = 0.018 \quad .$$

Thus to claim that A breaks down more often than B has only a very small risk. However, if we did not know that the data could be paired by month, then we would have done a two-sample $t$ test. It can be shown easily that the pooled sample variance is $s^2 = 8.5$ and the $p$-value for this test is

$$\Pr\left\{t_8 > \frac{2.8}{\sqrt{(8.5 \times 2/5)}}\right\} = \Pr\{t_8 > 1.52\} = 0.087 \quad .$$

The reliability of the conclusion from this test is not as good as the paired test because we had less information about how the data were collected.

## 5.3. Inference on the Population Variances

### *5.3.1 One sample*

a) Sample and assumptions: Same as (5.2.1a).

b) Parameter of interest: $\theta = \sigma^2$.

c) A reasonable estimate of $\theta$ : $S^2$ (see (5.2.1d)).

d) Known distribution of $\hat{\theta}$: $(n-1)S^2/\sigma^2$ has a chi-square distribution with $(n-1)$ degrees of freedom.

e) Remark: The distribution of $(n-1)S^2/\sigma^2$ is quite sensitive to the normality assumption.

### *5.3.2 Two samples*

a) Sample and assumptions: $Y_{11}, Y_{12}, \ldots, Y_{1n_1} \sim N(\mu_1, \sigma_1^2)$; $Y_{21}, Y_{22}, \ldots, Y_{2n_2} \sim N(\mu_2, \sigma_2^2)$.

b) Parameter of interest: $\theta = \sigma_1^2/\sigma_2^2$.

c) A reasonable estimate for $\theta$ : $\hat{\theta} = S_1^2/S_2^2$ (see computation of $S_1^2$ and $S_2^2$ in (5.2.2d).

d) Known distribution of $\hat{\theta}$ : $\hat{\theta}/\theta$ has an $F$-distribution with $n_1 - 1, n_2 - 1$

degrees of freedom, or

$$\frac{S_1^2/S_2^2}{\sigma_1^2/\sigma_2^2} \sim F_{n_2-1}^{n_1-1} \quad . \tag{5.3.2d}$$

*Example 5.3.1:* In Example 5.2.1, find a 90% confidence interval for the true variance $\sigma^2$.

*Solution:* Obviously, the confidence interval depends on that $(n-1)S^2/\sigma^2$ has a known chi-square distribution. Since $S^2$ is always positive, to bound $(n-1)S^2/\sigma^2$ from too small or too large numbers is the same as to bound it away from 0 and $\infty$. Thus we usually choose $\chi^2_{\nu,\alpha/2}$, $\chi^2_{\nu,1-\alpha/2}$ such that

$$\Pr\{\chi^2_{\nu,\alpha/2} \leqslant \nu S^2/\sigma^2 \leqslant \chi^2_{\nu.1-\alpha/2}\} = 1-\alpha \quad , \tag{5.3.3}$$

where $\nu = n-1$. In this particular example, $\nu = 9-1 = 8$, $\alpha = 0.1$, $S^2 = 24.445$. Hence a 90% confidence interval for $\sigma^2$ is

$$2.732 < 8 \times 24.445/\sigma^2 < 15.51 \quad ,$$

$$12.61 < \sigma^2 < 71.58, \quad \text{or} \quad 3.55 < \sigma < 8.46 \quad .$$

Note that this interval is rather large. In general, to have a good estimate of the variance from a small sample size is difficult.

*Example 5.3.2:* In Example 5.2.2, test the hypothesis that the two populations have the same variance.

*Solution:* Our hypotheses are $H_0 : \sigma_1^2 = \sigma_2^2$ versus $H_1 : \sigma_1^2 \neq \sigma_2^2$. The decision rule is to accept $H_1$ if $S_1^2/S_2^2$ is too different from 1, that is, when $S_1^2/S_2^2 > c_1$ or $S_2^2/S_1^2 > c_2$. At a given $\alpha$ level, $c_1$ and $c_2$ are chosen to be respectively $F^{n_1-1}_{n_2-1,1-\alpha/2}$ and $F^{n_2-1}_{n_1-1,1-\alpha/2}$. The $p$-value of this test is

$$\begin{aligned} p\text{-value} &= 2 \times \Pr\left\{F_{n_2-1}^{n_1-1} > s_1^2/s_2^2\right\} \qquad \text{when} \qquad s_1^2 > s_2^2 \\ &= 2 \times \Pr\left\{F_{n_1-1}^{n_2-1} > s_2^2/s_1^2\right\} \qquad \text{when} \qquad s_2^2 > s_1^2 \quad , \end{aligned}$$

where $s_1^2$ and $s_2^2$ are the two observed variances. In this example, the $p$-value is

$$\begin{aligned} p\text{-value} &= 2\Pr\{F_9^8 > 24.445/17.82\} \\ &= 2\Pr\{F_9^8 > 1.38\} = 0.3192 \quad . \end{aligned}$$

Obviously, we cannot confidently conclude that $\sigma_1^2 \neq \sigma_2^2$.

*Remark:* The one-sided two-sample variance test is much easier. For example, to test $H_0 : \sigma_1^2 = \sigma_2^2$ against $H_1 : \sigma_1^2 > \sigma_2^2$, the rejection rule is $S_1^2/S_2^2 > c$ and the $p$-value is $\Pr\{F_{n_2-1}^{n_1-1} > s_1^2/s_2^2\}$ .

## 5.4. Inference on Proportions

### *5.4.1 One sample*

a) Sample: Number of successes $X$ in $n$ observations. Let $p$ be the probability of success in each observation.

b) Parameter of interest: $\theta = p$.

c) A reasonable estimate of $\theta : \hat{\theta} = \hat{p} = X/n$.

d) Known distribution of $\hat{\theta}$: $n\hat{p} \sim \text{Bi}(n, p)$ and for large $n$, $n\hat{p} \sim N(np, npq), q = 1 - p$.

### *5.4.2 Two sample*

a) Sample: $X_1 \sim \text{Bi}(n_1, p_1), X_2 \sim \text{Bi}(n_2, p_2)$.

b) Parameter of interest: $\theta = p_1 - p_2$.

c) A reasonable estimate of $\theta$ : $\hat{\theta} = \hat{p}_1 - \hat{p}_2 = X_1/n_1 - X_2/n_2$.

d) Known distribution of $\hat{\theta}$ : For large $n_1$ and $n_2$ ,

$$\hat{p}_1 - \hat{p}_2 \sim N\left(p_1 - p_2, \frac{p_1 q_1}{n_1} + \frac{p_2 q_2}{n_2}\right), \quad q_i = 1 - p_i \ , \ i = 1, 2 \ .$$

Statistical inference on a one-sample proportion has been studied in Chapter 3. Here we will present the two-sample inference.

*Example 5.4.2:* In *Science* June 16, 1978, J. Levy and J. M. Levy reported that a right-handed male has a higher chance than a right-handed female of having a larger right foot. Their conclusion was based on 40 randomly selected right-handed males and 87 right-handed females. Their data is as follows:

| *Relative foot size* | *males* | *females* |
|---|---|---|
| left foot $\geqslant$ right | 12 | 73 |
| left foot $<$ right | 28 | 14 |

Do these data support their claim?

*Solution:* Let $p_1$ = probability that a right-handed male has a larger right foot, and $p_2$ = probability that a right-handed female has a larger right foot. Since this is the first experiment of this kind, the alternative hypothesis has to be $H_1 : p_1 \neq p_2$. Our decision rule must be $|\hat{p}_1 - \hat{p}_2| > c$. Thus the $p$-value of this test is

$$p\text{-value} = \Pr\{|\hat{p}_1 - \hat{p}_2| > c \mid H_0\}$$

$$= \Pr\left\{\frac{|\hat{p}_1 - \hat{p}_2 - 0|}{\sqrt{p_1 q_1/n_1 + p_2 q_2/n_2}} > \frac{c}{\sqrt{p_1 q_1/n_1 + p_2 q_2/n_2}}\right\} \qquad (5.4.1a)$$

where $c$ is the observed difference $|\hat{p}_1 - \hat{p}_2|$, or

$$p\text{-value} \doteq \Pr\left\{Z > |\hat{p}_1 - \hat{p}_2| \Big/ \sqrt{\frac{\hat{p}_1 \hat{q}_1}{n_1} + \frac{\hat{p}_2 \hat{q}_2}{n_2}}\right\}. \qquad (5.4.1b)$$

The last equality is an approximation because $p$'s and $q$'s are replaced by $\hat{p}$ and $\hat{q}$. Note also by the definition of the $p$-value that $\hat{p}_1$ and $\hat{p}_2$ in (5.4.1a) are two random variables, while the $\hat{p}$'s and $\hat{q}$'s in (5.4.1b) are their estimated values from this particular set of data. From the data

$$\hat{p}_1 = \frac{28}{40} = 0.7, \qquad \hat{p}_2 = \frac{14}{80} = 0.175\ ,$$

$$\frac{\hat{p}_1 \hat{q}_1}{n_1} + \frac{\hat{p}_2 \hat{q}_2}{n_2} = \frac{0.7 \times 0.3}{40} + \frac{0.175 \times 0.725}{80} = 0.0068\ .$$

Thus the $p$-value $= \Pr\{|Z| > |0.7 - 0.175/\sqrt{0.0068}\} = \Pr\{|Z| > 6.37\} \leqslant 10^{-4}$. Their conclusion should be very accurate.

## 5.5. Inference on Frequency Table and Histogram

Only hypothesis testing procedures will be dicussed in this section. As we have discussed in Remark 2 of Section 5.1, the confidence interval aspect of this type of inference is much more complicated. Interested readers can find the results of the estimation problem for a frequency table in Johnson & Koltz [12].

a) Sample: Observed frequencies in $k$ categories.

| | Category | | | |
|---|---|---|---|---|
| | 1 | 2 | $3 \ldots k$ | Total observations |
| Observed frequency | $o_1$ | $o_2$ | $o_3 \ldots o_k$ | $n = o_1 + o_2 + \ldots + o_k$ |

b) Hypotheses $H_0$: The data came from a given population with known distribution. $H_1 : H_0$ is not true.

c) The decision rule: The first step is to compute the expected frequencies in the $k$-categories under $H_0$. Denote them by $e_1, e_2, \ldots, e_k$ with $e_1 + e_2 + \ldots + e_k = n$, where the $e_i$'s do not have to be integers. Then $H_1$ is accepted if

$$Q = \sum_{i=1}^{k} \frac{(o_i - e_i)^2}{e_i} > c . \quad (5.5.1)$$

d) Known distribution of $Q$: For large $n$ and large $e_i$ (most $e_i \geqslant 5$), then $Q$ has an approximate chi-square distribution with $(k - 1 - r)$ degrees of freedom, where $r$ is the number of parameters estimated in order to calculate the $e_i$'s. Thus the relation between the significance level $\alpha$ and $c$ is

$$\alpha = \Pr\{\chi^2_{k-r-1} > c\} \quad , \quad \text{or} \quad c = \chi^2_{k-r-1,1-\alpha} \quad . \quad (5.5.2)$$

The $p$-value for given frequencies $o_1, o_2, \ldots, o_k$ is

$$p\text{-value} = \Pr\left\{\chi^2_{k-r-1} > \sum_{i=1}^{k} \frac{(o_i - e_i)^2}{e_i}\right\} \quad (5.5.3)$$

*Example 5.5.1:* A die was tossed 60 times with the following appearances:

| Faces | 1 | 2 | 3 | 4 | 5 | 6 | Total |
|---|---|---|---|---|---|---|---|
| Observed frequency | 8 | 5 | 20 | 15 | 6 | 6 | 60 |

Can we conclude that this is an unbalanced die?

*Solution:* Obviously, we wish to test

$H_0$ : The die is balanced, against

$H_1$ : The die is not balanced.

In order to make a decision, we first compute the expected frequencies under $H_0$ that the die is balanced. This is easy because we *expect* each face to appear 1/6 of the total 60 tosses, or $e_1 = e_2 = \ldots = e_6 = 10$. Thus by (5.5.3), the $p$-value of this test is

$$\Pr\left\{\chi^2_{k-r-1} > \sum_{i=1}^{6} \frac{(o_i - e_i)^2}{e_i}\right\} ,$$

where $o_i$'s are the observed frequencies 8, 5, 20, 15, 6, 6. A simple computation leads to

$$p\text{-value} = \Pr\{\chi^2_5 > 18.6\} = 0.0025 .$$

The degrees of freedom for the chi-square distribution is determined by $k = 6$, $r = 0$. We let $r = 0$ because no parameter was estimated.

*Example 5.5.2:* The following data were the last three digits (2,999, . . . km/sec) of Michelson's 100 measurements of the speed of light in km/sec. (*Astronomical Papers,* 1881, p. 231). Test the hypothesis that the data came from a normal distribution.

| Class | Interval | $o_i$ |
|---|---|---|
| 1 | (650, 700) | 2 |
| 2 | (700, 750) | 7 |
| 3 | (750, 800) | 15 |
| 4 | (800, 850) | 30 |
| 5 | (850, 900) | 23 |
| 6 | (900, 950) | 11 |
| 7 | (950, 1000) | 12 |
| | Total | 100 |

*Solution:* In order to find the expected frequencies under the null hypothesis that the data came from a normal distribution, we need to estimate the mean $\mu$ and variance $\sigma^2$ for a normal population. From the frequency table we have

$$\bar{X} \doteq (675 \times 2 + 725 \times 7 + 775 \times 15 + \ldots + 975 \times 12)/100$$

$$= 848$$

$$\sum_{i=1}^{100} X_i^2 \doteq 675^2 \times 2 + 725^2 \times 7 + 775^2 \times 15 + \ldots + 975^2 \times 12$$

$$= 72{,}447{,}500$$

or

$$S^2 = \frac{72{,}447{,}500 - 100 \times 848^2}{100 - 1}$$

$$= \frac{537{,}100}{99} = 5425$$

$$S = \sqrt{5425} = 74 \quad .$$

Thus $\mu$ is estimated to be 848 and $\sigma$ is estimated as 74. The probability of an observation falling in the first class is

$$\Pr\{650 < X < 700\} = \Pr\left\{\frac{650 - 848}{74} < Z < \frac{700 - 848}{74}\right\}$$

$$= \Pr\{-2.67 < Z < -2\}$$

$$= 0.02 \quad .$$

Thus the expected frequency in the first class $e_1 = 100 \times 0.02 = 2$. Similarly

$$e_2 = 100 \times \Pr\{700 < X < 750\}$$

$$= 100 \times \Pr\{-2 < z < -1.32\}$$

$$= 7.1 \quad .$$

Continuing in this manner, we can fill in the expected frequencies in the following table.

| class (i) | 1 | 2 | 3 | 4 | 5 | 6 | 7 | Total |
|---|---|---|---|---|---|---|---|---|
| $o_i$ | 2 | 7 | 15 | 30 | 23 | 11 | 12 | 100 |
| $e_i$ | 2 | 7 | 16 | 25 | 25 | 16 | 9 | 100 |
| $o_i - e_i$ | 0 | 0 | −1 | 5 | −2 | −5 | 3 | |

Thus

$$Q = \frac{0^2}{2} + \frac{0^2}{7} + \frac{(-1)^2}{16} + \ldots + \frac{3^2}{9} = 3.785 \ .$$

Because 2 parameters are estimated, the $p$-value for this test by (5.5.3) is

$$p\text{-value} = \Pr\{\chi^2_{7-2-1} > 3.785\} = \Pr\{\chi^2_4 > 3.785\} = 0.43 \ .$$

Hence we have no reason to reject the hypothesis that the experimental variation was normally distributed.

## 5.6. Correlation and Simple Linear Regression

Scientific investigation often has two goals. One is to correctly determine some characteristics of a population, such as for the mean $\mu > 25$ or for means $\mu_1 > \mu_2$. This has been a major topic in the previous five sections: to confidently describe some unknown quantity. In this chapter, we will discuss another goal of the scientific investigator, that is, to find the relationship between two, or more than two phenomena. We will restrict ourselves to two phenomena (or two variables). Variables can be related according to two types of models: deterministic and probabilistic. For example, we can state two well-known deterministic models: $E = mc^2$ denotes the relationship between energy $E$ and mass $m$, $s = \frac{1}{2}gt^2$ denotes the relationship between the displacement of a free falling body and the time $t$ it was released. If a relationship is probabilistic (or stochastic), we need statistical analysis. For example, let $X$ denote a person's height and $Y$ his weight. Then $X$ and $Y$ are apparently related, but this relationship is not deterministic. In other words, *given one's height, we cannot predict his weight with 100% accuracy, but we can predict his weight better than without his height.* In this case we say the pair $(X, Y)$ is a bivariate random variable. We say that $(X, Y)$ has a bivariate normal distribution if, roughly speaking, $X$ is normally distributed for any given $Y = y$ and vice versa. Bivariate normal distributions are again abundant in biological variations. The parameters that control a bivariate normal random variable $(X, Y)$ are

$$\mu_x = E(X) = \text{mean of } X, \quad \mu_y = E(Y) = \text{mean of } Y$$

$$\sigma_x^2 = E(X-\mu_x)^2 = \text{variance of } X,$$

$$\sigma_y^2 = E(Y-\mu_y)^2 = \text{variance of } Y,$$

$$\rho \equiv \rho_{xy} = E(X-\mu_x)(Y-\mu_y)/(\sigma_x \sigma_y)$$

$$= \text{correlation between } X \text{ and } Y. \tag{5.6.1}$$

It can be shown that $-1 \leqslant \rho \leqslant 1$ and that the density function of $(X, Y)$ is

$$f(x,y) = \frac{1}{2\pi\sigma_x\sigma_y\sqrt{1-\rho^2}} \exp\left\{\frac{-1}{2(1-\rho^2)}\left[\left(\frac{x-\mu_x}{\sigma_x}\right)^2 - 2\rho\,\frac{(x-\mu_x)(y-\mu_y)}{\sigma_x\sigma_y} + \left(\frac{y-\mu_y}{\sigma_y}\right)^2\right]\right\}. \tag{5.6.2}$$

It can also be shown that for a given value $Y = y$, $X$ has a normal distribution

$$X|_{Y=y} \sim N\left(\mu_x + \rho\,\frac{\sigma_x}{\sigma_y}(y-\mu_y)\,,\quad \sigma_x^2(1-\rho^2)\right). \tag{5.6.3}$$

Moreover, this distribution provides the best way (in terms of mean square error) to predict $X$ from the given information $Y = y$. From (5.6.3), it can be seen that when $\rho = 0$, then $X$, for the given $Y = y$, is distributed $N(\mu_x, \sigma_x^2)$. That is, the information that $Y = y$ does not affect the distribution of $X$. When this occurs, we say that $X$ and $Y$ are independent. Consider another extreme case, $\rho = 1$ or $\rho = -1$. Then the variance of $X$ given $Y = y$ is equal to $\sigma_x^2(1-1^2) = 0$. This means $X$ is totally predictable from $Y$. We say that $X$ and $Y$ are totally dependent on each other. In most practical cases, $-1 < \rho < 1$ and $X|_{Y=y}$ still has variation, but the variance has been reduced from $\sigma_x^2$ to $\sigma_x^2(1 - \rho^2)$ when $Y = y$ is known.

Though $(X, Y)$ is controlled by five parameters $\mu_x, \mu_y, \sigma_x^2, \sigma_y^2$, and $\rho$, all but the statistical inference for $\rho$ have been studied in Sections 5.2 and 5.3. For example, for a given sample $(X_1, Y_1), (X_2, Y_2), \ldots, (X_n, Y_n)$ from a bivariate normal distribution, the estimates for $\mu_x$, $\mu_y$, $\sigma_x^2$, and $\sigma_y^2$ are respectively,

$$\hat{\mu}_x = \overline{X} = \sum_{i=1}^{n} \frac{X_i}{n} ,$$

$$\hat{\mu}_y = \overline{Y} = \sum_{i=1}^{n} \frac{Y_i}{n} ,$$

$$\hat{\sigma}_x^2 = S_x^2 = \sum_{i=1}^{n} \frac{(X_i - \overline{X})^2}{(n-1)} ,$$

$$\sigma_y^2 = S_y^2 = \sum_{i=1}^{n} \frac{(Y_i - \overline{Y})^2}{(n-1)} .$$

By the definition of $\rho$, a reasonable estimate for $\rho$ is

$$\hat{\rho} = \frac{\dfrac{1}{n-1} \sum_{i=1}^{n} (X_i - \overline{X})(Y_i - \overline{Y})}{S_x S_y} . \qquad (5.6.4)$$

For large $n$,

$$\frac{1}{2} \ln \frac{1+\hat{\rho}}{1-\hat{\rho}} \sim N\left(\frac{1}{2} \ln \frac{1+\rho}{1-\rho} , \frac{1}{n-3}\right) , \qquad (5.6.5)$$

where $\rho$ is the true correlation.

*Example 5.6.1:* A study based on 1,000 pairs of fathers and sons showed that the average heights of fathers and sons are 68.5 inches with standard deviations 2.7 inches. The correlation was found to be 0.52.

(i) Find the reliability of estimated correlation 0.52, i.e. find a confidence interval for the true correlation $\rho$.

(ii) Assuming the father is 72 inches tall, what is the probability that his son will be taller? (Also assume the estimated values are close to the true values.)

*Solution:*

(i) Apparently we need to use (5.6.5) to construct a confidence interval for $\rho$. If we let $\theta = \frac{1}{2}\ln[(1+\rho)/(1-\rho)]$ and $\hat{\theta} = \frac{1}{2}\ln[(1+\hat{\rho})/(1-\hat{\rho})]$, then a 95% confidence interval for $\hat{\theta}$ is

$$\hat{\theta} \pm 1.96\sqrt{\frac{1}{n-3}} = \frac{1}{2}\ln\left(\frac{1+0.52}{1-0.52}\right) \pm 1.96\sqrt{\frac{1}{1000-3}}$$

$$= 0.576 \pm 0.062$$

$$= (0.514, 0.638) \quad .$$

Since $\ln[(1+\rho)/(1-\rho)]$ is a monotonic function of $\rho$, solving $\frac{1}{2}\ln(1+\rho)/(1-\rho) = 0.514$ and $0.638$, we have equivalently $[0.473, 0.577]$ is a 95% confidence interval for $\rho$.

(ii) Using (5.6.3) we find that the height of the son should be $X \sim N(68.5 + 0.52(72-68.5),\ 2.7^2(1-0.52^2)) = N(70.32,\ 5.32)$. Thus the probability of the son being taller than his father is

$$\Pr\{X > 72\} = \Pr\left\{Z > \frac{72-70.32}{5.32}\right\} = \Pr\{Z > 0.73\} = 0.23 \quad .$$

Notice that this probability is much less than the 50% chance we might expect. This is usually called the *correlation (regression) fallacy,* that is, although the two random variables are (highly) correlated, their extreme values do not have a similarly strong association.

The concept of simple linear regression is very similar to that of correlation except that one of the variables in the pair $(X, Y)$ is a fixed variable. For example, we may wish to find the relationship between $X$, the estimated execution time for a computer program by a computer programmer and $Y$, the actual execution time when the program is run. Here we see that $Y$ is a random variable because its value cannot be exactly determined before the program is run, but $X$ is a given number and hence not a random variable, and consequently we use small $x$ for $X$. The simplest relationship we can form between $x$ and $Y$ is

$$Y = \beta_0 + \beta_1 x + \epsilon \quad , \tag{5.6.6}$$

where $\beta_0$, $\beta_1$ are two unknown constants and $\epsilon$ is the error that is not predictable by this model. Model (5.6.6) is usually called a simple linear (regression) model. If we assume that $\epsilon$ has a normal distribution with 0 mean and an unknown variance $\sigma^2$, then given $n$ pairs of observations

$$(x_1, Y_1), (x_2, Y_2), \ldots, (x_n, Y_n) \quad ,$$

the minimum variance unbiased estimators for $\beta_0$ and $\beta_1$ are the $\hat{\beta}_0$ and $\hat{\beta}_1$ that minimize the error sum of squares $\sum_{i=1}^{n} (Y_i - \hat{\beta}_0 - \hat{\beta}_1 x_i)^2$. It can be shown by differentiation that

$$\hat{\beta}_1 = SS_{xY}/SS_x \quad , \qquad \hat{\beta}_0 = \overline{Y} - \hat{\beta}_1 \overline{x} \quad ,$$

where

$$\overline{x} = \sum_{i=1}^{n} \frac{x_i}{n} \quad , \qquad \overline{Y} = \sum_{i=1}^{n} \frac{Y_i}{n} \quad ,$$

$$SS_x = \sum_{i=1}^{n} (x_i - \overline{x})^2 = \sum_{i=1}^{n} x_i^2 - n\overline{x}^2 \quad ,$$

$$SS_Y = \sum_{i=1}^{n} (Y_i - \overline{Y})^2 = \sum_{i=1}^{n} Y_i^2 - n\overline{Y}^2 \quad ,$$

$$SS_{xY} = \sum_{i=1}^{n} (x_i - \overline{x})(Y_i - \overline{Y}) = \sum_{i=1}^{n} x_i Y_i - n\overline{x}\,\overline{Y} \quad . \tag{5.6.7}$$

Most scientists also call $\hat{\beta}_0$ and $\hat{\beta}_1$ the least squares estimators of $\beta_0$ and $\beta_1$. The minimized error sum of squares is usually denoted by

$$SSE = \sum_{i=1}^{n} (Y_i - \hat{\beta}_0 - \hat{\beta}_1 x_i)^2 = SS_Y - \hat{\beta}_1^2 \, SS_x \quad , \tag{5.6.8}$$

and it can be used to estimate the unknown variance $\sigma^2$ of $\epsilon$. More precisely

$$\hat{\sigma}^2 \equiv \text{MSE} = \text{SSE}/(n-2) \quad .$$

To use $\hat{\beta}_0$, $\hat{\beta}_1$, and $\hat{\sigma}^2$ to make inferences on $\beta_0$, $\beta_1$, and $\sigma^2$, we state their distributions as follows.

$$\frac{\hat{\beta}_1 - \beta_1}{\sqrt{\text{MSE}/\text{SS}_x}} \sim t_{n-2} \quad ,$$

$$\frac{\hat{\beta}_0 - \beta_0}{\sqrt{\Sigma x_i^2/n \cdot \text{MSE}/\text{SS}_x}} \sim t_{n-2} \quad ,$$

$$\text{SSE}/\sigma^2 \sim \chi^2_{n-2} \quad . \tag{5.6.9}$$

The estimated line $y = \hat{\beta}_0 + \hat{\beta}_1 x$ is usually called the regression line. It can be used to predict the future $y$ values for any given $x$. Obviously the point prediction for future $y$ given $x = x^*$ is

$$\hat{y} = \hat{\beta}_0 + \hat{\beta}_1 x^* \quad .$$

A $(1-\alpha)$ prediction interval for a new $Y$ to be observed at $x = x^*$ is

$$\hat{\beta}_0 + \hat{\beta}_1 x^* \pm t_{n-2,\,1-\alpha/2} \sqrt{\text{MSE}\left(1 + \frac{1}{n} + \frac{(x^* - \bar{x})^2}{\text{SS}_x}\right)} \quad . \tag{5.6.10}$$

*Example 5.6.2:* A list of execution times ($Y$) and their original estimated execution times ($x$) in seconds from 10 randomly selected large programs are given as follows:

| $x_i$: | 200 | 300 | 300 | 400 | 600 | 200 | 300 | 100 | 200 | 400 |
|---|---|---|---|---|---|---|---|---|---|---|
| $Y_i$: | 142 | 241 | 281 | 244 | 502 | 148 | 312 | 124 | 167 | 325 |

(i) Estimate the coefficients $\beta_0$ and $\beta_1$ if a simple linear model is used to express the relationship between $x$ and $Y$.

(ii) Can we claim that $\beta_1 < 1$, i.e. more people overestimate the execution time?

(iii) What is the predicted execution time of a program with 500 seconds estimated computing time?

*Solution:*

(i) First we display the data in a graph (see Fig. 5.6.1).

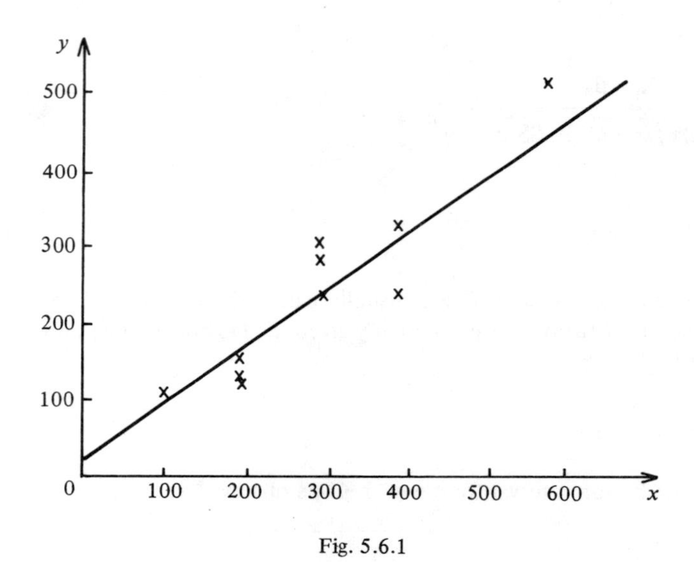


Fig. 5.6.1

Using (5.6.1) we have

$$\bar{x} = \sum_{i=1}^{10} \frac{x_i}{n} = 300 \quad , \qquad \bar{Y} = 248.6 \;,$$

$$SS_x = \sum_{i=1}^{10} x_i^2 - n\bar{x}^2 = 108 \times 10^4 - 10 \times 300^2 = 18 \times 10^4$$

$$SS_Y = \sum_{i=1}^{10} Y_i^2 - 10\bar{Y}^2 = 736884 - 618019.6 = 11.89 \times 10^4$$

$$SS_{xY} = \sum_{i=1}^{n} x_i Y_i - n\bar{x}\bar{Y} = 882800 - 745800 = 13.7 \times 10^4$$

$$\hat{\beta}_1 = SS_{xY}/SS_x = \frac{13.7}{18} = 0.76$$

$$\hat{\beta}_0 = \bar{Y} - \hat{\beta}_1 \bar{x} = 248.6 - 0.76 \times 300 = 20.6 \quad .$$

Thus the best fit line (the regression line) is $y = 20.6 + 0.76x$ which is shown in Fig. 5.6.1.

(ii) To confidently accept $H_1 : \beta_1 < 1$ over $H_0 : \beta_1 \geqslant 1$, we have the $p$-value of this test

$$p\text{-value} = \Pr\left\{t_{n-1} < \frac{\hat{\beta}_1 - 1}{\sqrt{\text{MSE}/SS_x}}\right\} = \Pr\left\{t_8 < \frac{0.76 - 1}{\sqrt{1.49/18}}\right\}$$

$$= \Pr\{t_8 < -0.83\} = 0.417 \quad .$$

Though the estimated $\beta_1$ is smaller than 1, it is too risky to claim $\beta_1 < 1$ based on the evidence we have.

(iii) The point prediction of $y$ at $x = 500$ is $\hat{y} = 20.6 + 0.76 \times 500 = 400.6$. Using (5.6.10) to construct a 95% confidence interval for $Y$, we have

$$400.6 \pm t_{8,\,0.975} \sqrt{0.187 \times 10^4 \left(1 + \frac{1}{10} + \frac{(500 - 300)^2}{18 \times 10^4}\right)}$$

$$= 400.6 \pm 2.306 \times 49.5 = 400.6 \pm 114.12 \quad .$$

**Exercise 5**

5.1 Show that the power of a one-sample $t$-test is invariant under the location and scale transformation, i.e., show that for a given significance level $\alpha$, to test

$$H_0 : \mu = \mu_0 , H_1 : \mu > \mu_0$$

for $N(\mu_1 , \sigma^2)$ has the same power as testing

$$H_0 : \mu = 0, \quad H_1 : \mu > 0$$

for $N((\mu_1 - \mu_0)/\sigma , 1)$. What is the significance of this result in simulation study?

5.2 Two methods, diffused based probe and ion implanted based probe, are generally used to manufacture IC devices. The percentage yields based on a sample size 20 on each method gave the following data:

$$x_1 = 50.3, \quad s_1 = 9.98, \quad x_2 = 59.0, \quad s_2 = 14.68 \quad ,$$

where $x_i$ and $s_i$ denote the sample means and sample standard deviations. Can you say that one method yields significantly (define) more than the other? State your assumptions and the reliability of your conclusion. (For data source, see Alvarez *et al.* (1984).)

5.3 A simulation study for the through-put time for a computer system produced the following data (in seconds)

$$421, 314, 43, 72, \ldots, 198 \quad .$$

The sample size, mean, and standard deviation are respectively 225, 321, 60. Find a 95% confidence interval for the mean through-put time.

5.4 In the previous exercise, 135 outputs took more than 300 seconds. Find a 99% confidence interval for the proportion of programs that require more than 300 seconds of through-put time.

5.5 We would like to test the uniformity and independence of a random number generator. One thousand random numbers are generated and we found,

(i) 52% of them are above 0.5, and

(ii) the sample correlation between 500 consecutive pairs is 0.07. Complete the tests.

5.6 (*Programming exercise*) Write down a subroutine that will find the $p$-value for one-sample and two-sample $t$ tests for the means and the chi-square, $F$ tests for the variances. You need to combine this exercise and the subroutines in Ex. 2.15. Test your program with the following data.

Data from population 1

10, 12, 18, 24, 26, 27, 27, 30, 34, 36, 37

Data from population 2

12, 13, 15, 18, 19, 20, 20, 22, 24, 25, 25, 26, 26

For $H_0 : \mu_1 = 25, H_1 : \mu_1 \neq 25, p = 0.82$;

For $H_0 : \mu_1 = \mu_2 , H_1 : \mu_1 > \mu_2 , p = 0.51$;

For $H_0 : \sigma_1 \geqslant 8, H_1 ; \sigma_1 < 8, p = 0.767$;

For $H_0 : \sigma_1^2 = \sigma_2^2 , H_1 : \sigma_1^2 > \sigma_2^2 , p = 0.0223$.

5.7 (*Programming exercise*) The purpose of this exercise is to check the validity and robustness of the $t$ and chi-square confidence intervals for the mean and variance. Let $\bar{x}$ and $s^2$ be the sample mean and variance from a sample of size $n$. The $(1 - \alpha)$ confidence intervals for the population mean $\mu$ and variance $\sigma^2$ are respectively.

(a) $\bar{x} \pm t_{n-1,\, \alpha/2}\, s/\sqrt{n}$ , and

(b) $[(n-1)s^2/\chi^2_{n-1,1-\alpha/2} ,\ (n-1)s^2/\chi^2_{n-1,\,\alpha/2}]$ .

Check their validity by the definition of a confidence interval with 1,000 simulations for each of the following cases. (Use $\alpha = 0.10$ and $n = 10$.)

(i) The population is $N(0.5, 2)$.

(ii) The population is uniform in $(1, 3)$.

(iii) The population has a density function

$$f(x) = 6x(1-x), \qquad 0 < x < 1 \ .$$

Your conclusion should be stated as "There is (or is not) enough evidence to claim the invalidity of (A) and/or (B) for this population." The reliability of your conclusion is important.

5.8 The following frequency table summarizes our findings on the lifetime of a brand of vacuum tubes. Test the null hypothesis that their lifetime is normally distributed.

| | Class intervals (in 100 hours) | | | | | | | |
|---|---|---|---|---|---|---|---|---|
| | 0-10 | 10-11 | 11-12 | 12-13 | 13-14 | 14-15 | 15-16 | 16-20 |
| Observed Frequency | 6 | 10 | 12 | 14 | 20 | 45 | 25 | 12 |

5.9 A computer company has sold 1,000 microcomputers MX421 in a city and it is responsible for their repairs during the 2-year warranty period. The following observations are recorded.

| | Number of repairs | | | | | | | |
|---|---|---|---|---|---|---|---|---|
| | 0 | 1 | 2 | 3 | 4 | 5 | 6 | >6 |
| No. of weeks with this many repairs | 23 | 28 | 15 | 10 | 12 | 8 | 4 | 0 |

Test the hypothesis that the weekly repairs are Poisson distributed and discuss the practical significance of this test.

5.10 A well-known brand of washing machine is sold in five colors and the market researcher wants to study their popularity. The following frequency table is obtained.

| Avocado | Tan | Red | Blue | White | Total sales |
|---|---|---|---|---|---|
| 88 | 65 | 52 | 40 | 55 | 300 |

Does the researcher have enough evidence to reject the hypothesis that the sales are independent of colors?

5.11 It is suspected that the breakdown rate of a computer is linearly related to the room temperature. The following data were obtained.

| Temp. °F | 55 | 57 | 59 | 62 | 64 | 65 | 67 | 72 | 73 | 75 | 78 | 81 |
|---|---|---|---|---|---|---|---|---|---|---|---|---|
| Breakdowns (hrs/week) | 3.4 | 3.6 | 2.2 | 2.5 | 4.0 | 5.3 | 7.2 | 8.5 | 7.2 | 6.3 | 8.0 | 7.7 |

Is there enough evidence to support the suspicion at this temperature range?

5.12 The grade point average (GPA) is often used by schools to accept new students and by companies to hire new employees. The following data were observed in a computer science department.

| High school GPA | University GPA | Starting Salary |
|---|---|---|
| 3.2 | 2.5 | 18,000 |
| 3.6 | 3.1 | 21,000 |
| 2.8 | 2.6 | 21,000 |
| 2.5 | 2.7 | 19,000 |
| 3.5 | 3.5 | 22,000 |
| 3.3 | 3.9 | 25,000 |
| 2.9 | 2.9 | 18,500 |
| 3.4 | 3.3 | 20,000 |
| 3.7 | 3.2 | 19,500 |
| 3.0 | 2.8 | 20,500 |

Test the correlation between the university GPA and the high school GPA, and that between the starting salary and university GPA. Can you claim the correlations are positive?

5.13 Two word-processing softwares are to be compared. Each was taught to 10 secretaries each. After the classes ended, a manuscript was given to the secretaries and here are the times (in minutes) they took to finish their work.

| | | | | | | | | | | |
|---|---|---|---|---|---|---|---|---|---|---|
| Processor 1 | 56 | 67 | 87 | 67 | 85 | 67 | 90 | 87 | 77 | 86 |
| Processor 2 | 56 | 63 | 56 | 78 | 76 | 53 | 77 | 82 | 85 | 64 |

Can we claim that one of the pieces of software is more time-saving than the other?

5.14 In a computing center, the old utility frequency table for the following languages are

| FORTRAN | BASIC | PASCAL | C | APL | PL-1 |
|---|---|---|---|---|---|
| 30% | 20% | 20% | 15% | 10% | 5% |

A recent record showed the usage frequencies of 200 calls are

| FORTRAN | BASIC | PASCAL | C | APL | PL-1 |
|---|---|---|---|---|---|
| 50 | 42 | 40 | 33 | 21 | 14 |

Can we claim that the utility frequency has changed?

5.15 Two searching algorithms are compared by 10 randomly selected inputs. The time (in seconds) to finish the searching tasks are recorded in the following:

| Job | 1 | 2 | 3 | 4 | 5 | 6 | 7 | 8 | 9 | 10 |
|---|---|---|---|---|---|---|---|---|---|---|
| Arg. 1 | 26 | 10 | 15 | 16 | 32 | 8 | 22 | 30 | 13 | 13 |
| Arg. 2 | 24 | 10 | 17 | 17 | 30 | 7 | 23 | 28 | 13 | 11 |

Can we conclude that one algorithm is faster than the other?

# Chapter 6
# PERMUTATION TESTS

## 6.1. Knowing the Distribution of a Statistic

Many of the statistics and tests that we have seen thus far are most useful when the population in question has certain specific characteristics. For example, a $t$-statistic is most appropriate when we wish to compare sample means from two normal populations having the same unknown variance $\sigma^2$. A test can be conducted by comparing the $t$-statistic with extreme values from the $t$-distribution. If our statistic is more extreme than the critical values of the $t$-distribution, then we are able to conclude that the null hypothesis $H_0 : \mu_1 = \mu_2$ is false. Furthermore, the critical values from the $t$-distribution provide us with the significance level of the test: i.e., the probability of falsely rejecting $H_0$.

In cases such as this, the distribution of the test statistic is known (assuming $H_0$ is true). We are able to develop a valid test of $H_0$ by using a decision rule which rejects $H_0$ (accept $H_1$) whenever an extreme value of the statistic is observed. Usually the test is constructed so that the probability of falsely rejecting $H_0$, when $H_0$ is really true, is .05. This five percent probability is very dependent on the distribution of the test statistic. For example, if we assume that the test statistic follows a normal distribution with a particular mean and variance, then we can find critical values for our test statistic by using tables of the normal distribution. These critical values will be chosen for a particular significance level (say, five percent) so that our test statistic will have only a five percent chance of being outside the critical values when the null hypotheses is true.

But what if our test statistic doesn't actually follow a normal distribution? Then our critical values will probably not be determining an $\alpha$-level of .05, but rather some other $\alpha$-level, which we cannot know without knowing the exact distribution of the test statistic.

To get around the problem of being uncertain about the distribution of a test statistic, another procedure has developed over the years. We desire a procedure which is "distribution-free"; that is, the distribution of the test statistic should not depend on the type of population which we happen to be sampling from. Ideally, a distribution-free test statistic would behave identically (have the same distribution) under any conditions, regardless of the shape of the underlying population. The "permutation test", which we introduce in the remaining sections of this chapter, gives a statistic this distribution-free property. R. A. Fisher was among the first, in the early part of this century, to propose such a test (also called a "randomization" test) as a method of controlling the distribution of a test statistic.

We will discuss three examples in the following sections of this chapter. Each example illustrates an hypothesis testing problem, for which a permutation test may be used. Other examples are indicated in the exercises at the end of the chapter.

## 6.2. A Permutation Test for Two Independent Samples

We begin this section with an example of a permutation test.

*Example 6.2.1:* Suppose that six independent measurements are taken in an experiment where we wish to compare two experimental treatments. Let us call the two treatments A and B. Three measurements are taken from treatment A. Suppose we observe the values 17, 25, and 30 from this treatment. The other three measurements are taken from treatment B. Let us suppose that these values turn out to be 15, 20, and 21. We want to know whether these values provide evidence that treatment A has a higher population mean than treatment B. How can we know?

Statistic textbooks would tell us to compare the sample means:

$$\overline{X}_{\mathrm{A}} = (17 + 25 + 30)/3 = 24.0$$

$$\overline{X}_{\mathrm{B}} = (15 + 20 + 21)/3 = 18.7 \quad .$$

$\overline{X}_{\mathrm{A}}$ appears to be quite a bit higher than $\overline{X}_{\mathrm{B}}$. In fact, $\overline{X}_{\mathrm{A}} - \overline{X}_{\mathrm{B}} = 5.3$. But how can we judge the statistical significance of the difference? We need a statistical test of the

hypothesis $H_0 : \mu_{\mathrm{A}} = \mu_{\mathrm{B}}$

versus $H_1 : \mu_{\mathrm{A}} > \mu_{\mathrm{B}}$ .

If we knew the populations for treatment A and treatment B were both normal, we could compute a $t$-statistic. But who can tell about normality when all we have is three values from each population? A permutation test can be used in this situation with no assumptions at all about populations A and B.

The permutation test begins by pooling the six values together into one group. We then consider all possible sets of three values for treatment A from the six possibilities. If we choose three of the values to belong to group A, then the other three must belong to treatment B. How many possible ways can this be done? According to some counting rules used in the binomial theorem, six objects can be divided into two sets of three in $\binom{6}{3} = \frac{6!}{3!3!} = 20$ different ways. Table 6.2.1 lists the 20 different possible results for $\overline{X}_A - \overline{X}_B$, if we had started with the six values 15, 17, 20, 21, 25, and 30. The 20 different results are ordered according to the resulting values of $\overline{X}_A - \overline{X}_B$. How many of the 20 different groupings have a difference $\overline{X}_A - \overline{X}_B$ as large as 5.3? Examination of the table reveals that in only three of the 20 listed possibilities is $\overline{X}_A$ larger than $\overline{X}_B$ by as much as 5.3. We conclude then, from the permutation test, that the significance of our actual observed value of $\overline{X}_A - \overline{X}_B = 5.3$, is, 3 out of 20, i.e. an observed significance level of .15. If we began our experiment hoping to reject $H_0$ at the 5% level of significance, then we have failed to do so. We would only be able to reject $H_0$ at the 15% level of significance, that is, with 85% confidence. If this is not conclusive enough evidence, we must fail to reject $H_0$.

What are the characteristics of the test we have just seen? The main difference between this test and the usual two sample $t$-test, is that this test, like nearly all permutation tests, assumes that the data provided from the experiment are fixed values. Hence we only examine different "permutations" of the given values between the two treatments. If the treatments are equivalent in terms of their effect on observed data, then each observed value can be considered to have been equally likely to have come from either treatment A or treatment B. The different arrangements of the given values are then equally likely. In this example, there were 20 such rearrangements. In general, with $n_A$ observations from treatment A and $n_B$ observations from treatment B, there are

$$\binom{n_A + n_B}{n_A} = \frac{(n_A + n_B)!}{n_A!\,n_B!}$$

different rearrangements of the values between the two treatments. In assessing the significance of our observed values, we rank the arrangements according to

Table 6.2.1

| Treatment A | Treatment B | $\bar{X}_A - \bar{X}_B$ |
|---|---|---|
| 21, 25, 30 | 15, 17, 20 | 8.0 |
| 20, 25, 30 | 15, 17, 21 | 7.3 |
| 17, 25, 30 | 15, 20, 21 | 5.3 |
| 20, 21, 30 | 15, 17, 25 | 4.7 |
| 15, 25, 30 | 17, 20, 21 | 4.0 |
| 17, 21, 30 | 15, 20, 25 | 2.7 |
| 17, 20, 30 | 15, 21, 25 | 2.0 |
| 15, 21, 30 | 17, 20, 25 | 1.3 |
| 20, 21, 25 | 15, 17, 30 | 1.3 |
| 15, 20, 30 | 17, 21, 25 | 0.7 |
| 17, 21, 25 | 15, 20, 30 | −0.7 |
| 15, 17, 30 | 20, 21, 25 | −1.3 |
| 15, 21, 25 | 17, 20, 30 | −2.0 |
| 15, 20, 25 | 17, 21, 30 | −2.7 |
| 17, 20, 21 | 15, 25, 30 | −4.0 |
| 15, 17, 25 | 20, 21, 30 | −4.7 |
| 15, 20, 21 | 17, 25, 30 | −5.3 |
| 15, 17, 21 | 20, 25, 30 | −7.3 |
| 15, 17, 20 | 21, 25, 30 | −8.0 |

their value of our statistic of interest. We assign a probability of

$$1 \Big/ \binom{n_A + n_B}{n_A} = \frac{n_A! n_B!}{(n_A + n_B)!}$$

to each of the arrangements, and count how many of these ranks are as high or higher than our observed value. This provides an observed significance level (also called a $p$-value) for our statistic, from which we may draw our conclusion. If the $p$-value is not sufficiently low, we do not reach the desired level of confidence in claiming that treatment A yields a higher average than treatment B. In such a case, our experiment has failed to give us conclusive evidence in favor of the alternative hypothesis $H_1$.

*Remark:* The procedure is easily adapted to a test with a two-sided alternative

hypothesis: $H_0 : \mu_A = \mu_B$

versus: $H_1 : \mu_A \neq \mu_B$ .

In this case, we would rank the different arrangements according to their values of $|\bar{X}_A - \bar{X}_B|$. In the previous example, this would change the observed significance level of our statistic from .15 (3 out of 20) to .30 (6 out of 20); that is, 6 out of the 20 arrangements would have values of $|\bar{X}_A - \bar{X}_B|$ as large or larger than our observed value of $|24.0 - 18.7| = 5.3$.

*Example 6.2.2:* The permutation procedure is not so easily carried out as the sample sizes within the two treatments grow larger. Suppose for example that we observe the data in Table 6.2.2.

Table 6.2.2

| | | | | | | | | | | |
|---|---|---|---|---|---|---|---|---|---|---|
| Treatment A: | 23 | 26 | 17 | 19 | 30 | 24 | 35 | 31 | 26 | 21 |
| Treatment B: | 21 | 14 | 19 | 25 | 26 | 12 | 20 | 16 | 13 | 15 |

Then we have

$$n_A = n_B = 10$$

and

$$\bar{X}_A - \bar{X}_B = 25.2 - 18.1 = 7.1 \quad .$$

The number of different possible arrangements of values in this design becomes prohibitively large:

$$\binom{20}{10} \doteq \frac{20!}{10!10!}$$

$$= 184{,}756 \quad .$$

In a case such as this, the computer becomes a great aid. We can easily generate a random sample of 999 of the arangements of the 20 observations through a computer simulation. This simulation would proceed as follows: We generate an arrangement of the 20 values by storing the given values in an array: $X(1)$ $X(2) \ldots X(19)$ $X(20)$. We then randomly generate a grouping of 10 of the 20 values to be placed under treatment A. The remainder are placed in treatment B. The statistic $\bar{X}_A - \bar{X}_B$ is then computed for that particular sample grouping.

What is the best way to generate a random grouping for treatment A? One method would be to generate 10 random integers from 1 to 20, being careful not to allow repetition. These ten integers would correspond to the elements in the array being grouped with treatment A. Another method would be to assign binary values 0 and 1 to each element of the array. The ten zero's would correspond to the ten elements in treatment A, while the ten one's would correspond to the elements in treatment B. For instance, the algorithm might look like this:

```
(1) DIMENSION ARRAY T(20)
(2) INITIALIZE ARRAY T = 0
     COUNTER = 1
     NUMBER-OF-ZEROS = 0
     NUMBER-OF-ONES = 0
(3) START LOOP TO ASSIGN VALUES OF 0 AND 1 TO EACH MEMBER OF
    ARRAY T:
     GENERATE UNIFORM RANDOM NUMBER U
     SET B = INT(2*U)
     SET T(COUNTER) = B
     IF B = 0
       THEN ADD 1 TO NUMBER-OF-ZEROS
```

```
        ELSE ADD 1 TO NUMBER-OF-ONES
      ADD 1 TO COUNTER
      IF NUMBER-OF-ZEROS = 10
        THEN BRANCH TO (4)
        ELSE IF NUMBER-OF-ONES = 10
          THEN BRANCH TO (5)
          ELSE GO BACK TO (3)
(4) SET EACH REMAINING VALUE OF ARRAY T = 1
    STOP
(5) SET EACH REMAINING VALUE OF ARRAY T = 0
    STOP
```

Once a random grouping is generated, and the statistic is computed, the new value for $\overline{X}_A - \overline{X}_B$ can be compared to the value computed from the actual experiment (7.1 in this example). We wish to know what proportion of the randomly generated sample of groupings yield a value of $\overline{X}_A - \overline{X}_B$ as large or larger than 7.1. The percentile rank of the actual statistic (7.1) is that proportion of the groupings for which $\overline{X}_A - \overline{X}_B$ is less than 7.1; i.e.

$$\text{Percentile Rank} = \frac{(\text{number of groupings with } \overline{X}_A - \overline{X}_B < 7.1)}{1000} .$$

For example, if 957 out of the 999 random groupings yield a value of $\overline{X}_A - \overline{X}_B$ less than the original observed value of 7.1, then the percentile rank of our statistic would be .957, or 95.7%. The attained significance, assuming we are using only the upper tail of the distribution to reject $H_0$, would be $1 - .957 = .043$, or 4.3%. Thus, if this were our result, we would be justified in rejecting $H_0: \mu_A = \mu_B$ in favor of the alternative $H_1: \mu_A > \mu_B$ with an $\alpha$-level of .05.

## 6.3. A Permutation Test for a Paired Two-Sample Design

The discussion in Section 6.2 proceeded under the assumption that the two samples for treatment A and treatment B were drawn from independent populations. However, it is also possible to develop a permutation test from a paired comparison type of experiment; that is, when each observation from treatment A is paired with a corresponding observation from treatment B. In such a case, the pairs of values are fixed within each arrangement of the values. The different arrangements are derived by merely exchanging the two values within a given pair. When this is done in all possible ways, there are $2^n$ different possible arrangements of $n$ distinct pairs of values. (Why?) Thus each of the rearrangements will be assigned a probability of $1/2^n$. The significance of the observed statistic can then be assessed by ranking the $2^n$ different values, using either a one-sided procedure or a two-sided procedure. In Example 6.2.1, suppose the six values were observed in 3 pairs, such as (17, 15), (25, 20) and (30, 21), with the first member of each pair from treatment A and the second member from treatment B. Then there are $2^3 = 8$ different values of $\overline{X}_A - \overline{X}_B$ which can be derived from these pairs. These 8 values are shown in order, in Table 6.3.1. The significance level of our observed statistic is seen to be 1 out of 8, or .125, since our observed value of $\overline{X}_A - \overline{X}_B = 5.3$ ranks first among the 8 values.

Table 6.3.1

| | Pair 1 | Pair 2 | Pair 3 | $\overline{X}_A - \overline{X}_B$ |
|---|---|---|---|---|
| Treatment A | 17 | 25 | 30 | 5.3 |
| Treatment B | 15 | 20 | 21 | |
| Treatment A | 15 | 25 | 30 | 4.0 |
| Treatment B | 17 | 20 | 21 | |
| Treatment A | 17 | 20 | 30 | 2.0 |
| Treatment B | 15 | 25 | 21 | |
| Treatment A | 15 | 20 | 30 | 0.7 |
| Treatment B | 17 | 25 | 21 | |
| Treatment A | 17 | 25 | 21 | −0.7 |
| Treatment B | 15 | 20 | 30 | |
| Treatment A | 15 | 25 | 21 | −2.0 |
| Treatment B | 17 | 20 | 30 | |
| Treatment A | 17 | 20 | 21 | −4.0 |
| Treatment B | 15 | 25 | 30 | |
| Treatment A | 15 | 20 | 21 | −5.3 |
| Treatment B | 17 | 25 | 30 | |

As was done in Example 6.2.2, a larger value for $n$ (the number of pairs), can be handled through the use of a simulation experiment. In such a situation, we would begin with the data stored in two arrays:

$$X(1)X(2)\ldots X(n-1)X(n)$$

and

$$Y(1)Y(2)\ldots Y(n-1)Y(n) \quad .$$

We would generate a random rearrangement of the values in $X$ and $Y$ by randomly switching approximately half of the $(X, Y)$ pairs. For instance, we might use the following algorithm:

```
(1) DIMENSION ARRAYS X(n), Y(n), A(n), B(n).
(2) STORE ACTUAL DATA IN ARRAYS X AND Y.
(3) INITIALIZE A-TOTAL = 0 AND B-TOTAL = 0.
(4) DO THE FOLLOWING FOR EACH I FROM 1 TO n:
      GENERATE A UNIFORM RANDOM NUMBER U
      IF U > .5
           THEN SET A(I) = X(I)
                AND B(I) = Y(I)
           ELSE SET A(I) = Y(I)
                AND B(I) = X(I)
      ADD A(I) TO A-TOTAL.
      ADD B(I) TO B-TOTAL.
(5) COMPUTE X̄_A = A-TOTAL/n
      AND X̄_B = B-TOTAL/n.
    STATISTIC = X̄_A − X̄_B.
```

This can be repeated 999 times (or as many times as desired) to get an estimate of the percentile rank of the statistic that was actually observed.

## 6.4. A Permutation Test for a Contingency Table

Permutation tests sometimes can be done without doing the actual permutation. Often we observe data in the form of a two-way table, where each entry in the table represents the number of observations which were observed in that cell.

For example, suppose we have twenty laboratory mice with similar cancerous tumors. We wish to test the effect of a new experimental drug on the tumors. We separate the mice into two groups. One group of mice we use as a control group, and administer to them only a placebo. Each mouse in the other group receives the new drug treatment. We observe what change occurs in the size of the tumor: either reduction or enlargement. We can record the results as in Table 6.4.1 below:

Table 6.4.1

| | Placebo | New Drug | Totals |
|---|---|---|---|
| Reduced Tumor | 1 | 7 | 8 |
| Enlarged Tumor | 9 | 3 | 12 |
| Totals | 10 | 10 | 20 |

We now ask what is the significance level of our table? To test a null hypothesis such as

$H_0$ : The drug has no effect

versus $H_1$ : The drug helps to reduce tumor size,

we must measure the probability that data like this can occur by chance, when the drug really has no effect on the size of a tumor. We must determine the probability that, of the 8 laboratory mice who had tumors reduced in size, at least 7 would happen to be from the drug group, by chance alone. That is, if we are assuming that the drug really has no effect, presumably we would still have seen 8 reductions in tumor size and 12 enlargements, regardless of how many mice received the new drug treatment. Thus we need only find the probability of randomly distributing the 20 mice between the two groups of 10 mice, so that 7 or 8 of the mice with reductions fall in the group receiving the new drug treatment.

This probability can be found using a formula involving binomial coefficients. This type of probability formula is called a hypergeometric probability (see Section 1.7).

P(exactly 7 mice in the experimental group have a reduction in tumor size)

$$= \frac{\binom{8}{1}\binom{12}{9}}{\binom{20}{10}}$$

$$= \frac{8!\,12!\,10!\,10!}{20!\,1!\,7!\,9!\,3!} = .0004$$

and

P(exactly 8 mice in the experimental group have a reduction in tumor size)

$$= \frac{\binom{8}{0}\binom{12}{10}}{\binom{20}{10}}$$

$$= \frac{8!\,12!\,10!\,10!}{20!\,0!\,8!\,10!\,2!} = .0095.$$

The total of these two probabilities is then .0099. This is the attained significance level of our observed data.

**Exercise 6** (*all programming exercises*)

6.1 Perform the simulation indicated in Section 6.2 to determine the significance level of the data given in Table 6.2.2. Create 999 random groupings of the given values to determine the percentile rank of the statistic $\overline{X}_A - \overline{X}_B = 7.1$.

6.2 Two computers are to be compared by running a series of 12 benchmark programs. The times (in seconds) required for running each program are recorded for each computer. These times are then compared to determine which computer will operate faster. The results are listed below:

| Computer | Program # | | | | | | | | | | | |
|---|---|---|---|---|---|---|---|---|---|---|---|---|
| | 1 | 2 | 3 | 4 | 5 | 6 | 7 | 8 | 9 | 10 | 11 | 12 |
| A | 44 | 55 | 71 | 35 | 98 | 89 | 156 | 74 | 102 | 61 | 246 | 196 |
| B | 56 | 61 | 98 | 30 | 95 | 97 | 180 | 73 | 129 | 68 | 288 | 231 |

(a) Test to see if there is a significant difference between the two computers, by generating 199 additional sets of permutations of the 12 pairs. Compute the absolute value of the difference between the two means for each of the 199 permutations that you generate. Approximate the significance level of the actual difference between the means.

(b) Compute the exact significance level by generating all possible permutations of the 12 pairs of values. (How many permutations are there all together?) For each permutation, compute the absolute value of the difference between the two means, as in part (a). Compare the actual difference in means with that computed from each of the generated permutations, in order to find the exact significance level.

6.3 A person claims to be able to discriminate between the taste of the New Coke and the taste of Coke Classic. To test this claim, the person is given a sequence of unmarked drinks, one per day, for a total of 30 days. 15 of the drinks were the New Coke, and the other 15 were the Coke Classic. After tasting each drink, the person is asked to identify which type of Coke it is. The results are given below, showing how many times the person was able to correctly identify each type of drink:

| | Actual drink was | | |
|---|---|---|---|
| Identified as | New Coke | Coke Classic | Totals |
| New Coke | 11 | 4 | 15 |
| Coke Classic | 4 | 11 | 15 |
| Totals | 15 | 15 | 30 |

(a) Approximate the significance level of this table, by randomly generating 499 other tables with the same marginal totals. Compare the number of correct classifications in each generated table with the number of correct classifications in the given table.

(b) Compute the exact significance level of the table, by computing the exact probability of having 11, 12, 13, 14 or 15 correct identifications of each type of Coke. Use the hypergeometric probability formula.

6.4 To determine if there is any correlation between the adult height of U.S. males and the height of their fathers, a survey was taken of 10 adult males. The survey ascertained the height of the respondent and the respondent's father. The results are shown below:

| Respondent Number | 1 | 2 | 3 | 4 | 5 | 6 | 7 | 8 | 9 | 10 |
|---|---|---|---|---|---|---|---|---|---|---|
| Respondent's Height | 68 | 71 | 70 | 69 | 70 | 75 | 72 | 67 | 73 | 71 |
| Father's Height | 63 | 69 | 70 | 68 | 66 | 73 | 71 | 67 | 74 | 70 |

We wish to test for a significant positive correlation in this data. Assume we construct a null hypothesis of no correlation between the heights. Then if $H_0$ were true, we might just as easily observe a completely different pairing of the heights of the fathers with the heights of the respondents. Determine an approximate significance level for the correlation coefficient of this data set by generating 199 other arrangements of these values, and calculating the correlation coefficient for each arrangement. (To generate a new arrangement of these values, keep the first two rows fixed. Generate a random ordering for the third row, thus changing the pairings of heights of fathers and sons.)

## Chapter 7

# JACKKNIFE AND BOOTSTRAP METHODS

### 7.1. Estimating MSE from Only One Sample

In the science of statistics, estimators are proposed and used for a variety of purposes. Some estimators are based on statistical theory and particular assumptions about the distribution of the random variables being observed. Such estimators often are said to be parametric in nature, as they estimate an unknown parameter of the assumed distribution. Examples of these might be certain maximum likelihood estimators or minimum variance unbiased estimators, as discussed in Chapter 4.

Other statistics make fewer assumptions about the form of the underlying distribution, but estimate some general characteristic of the population. These statistics might be called nonparametric estimators, because of their use of fewer specific distributional assumptions. Examples of these more general estimators include statistics such as the sample mean, the sample median, the Pearson product-moment correlation coefficient, the sample variance and standard deviation, and sample percentile estimators (also called quantiles). These might also be considered parametric in some cases, as they can often be derived using maximum likelihood or minimum variance unbiased procedures.

Of course, the search for better statistics continues, particularly in the more general nonparametric setting, where there are fewer restrictions on what kind of statistics can be considered. One criterion used to judge the quality of new estimators has been discussed in Chapter 4, namely the mean square error criterion (MSE). Recall that the MSE of an estimator $\theta$ is made up of two components:

(1) the variance $\mathrm{Var}(\hat{\theta}) = E[(\hat{\theta} - E(\hat{\theta}))^2]$ and

(2) the bias $B(\hat{\theta}) = E(\hat{\theta}) - \theta$.

It is stated in Chapter 4, that

$$\text{MSE} = \text{Var}(\hat{\theta}) + B^2(\hat{\theta}) \quad .$$

Thus, variance and bias are two major qualities to be studied for any new statistic that is proposed. When an estimator $\hat{\theta}$ is developed with smaller variance and/or smaller absolute bias ($|B(\hat{\theta})|$), then it may become a viable competitor to existing estimators. Of course, if the variance and bias remain small for a wide variety of possible underlying distributional assumptions, then the statistic becomes very attractive because of its robust behavior. This is a goal of non-parametric estimation procedures.

With the advent of computers and their ability to do massive amounts of calculations in a short period of time, new opportunities have been opened for estimators which might otherwise have been overlooked for being too complex. Many statistics can now be studied using simulation, for example to determine the variance and bias, where before the computer age, analysis was not feasible.

Two new techniques for studying new estimators have most recently come to the forefront in statistical research. The jackknife technique was originally proposed by Quenouille in 1956, with additional insight provided by Tukey in 1958. It has been developed since that time for a wide variety of areas and uses. The general effect of using the jackknife technique on an existing estimator is that it decreases the bias of the estimator, often without significantly changing the variance of the original estimator. In addition, the jackknife technique provides an estimate of the variance of the original estimator. This is important in studying the MSE of the estimator. It is also crucial if the estimator is to be used in forming confidence intervals for the true unknown parameter. These qualities have made the jackknife a popular tool among today's statisticians.

Another tool was developed even more recently, primarily by Brad Efron of Stanford University, beginning in the late 1970's. This method is called the bootstrap technique. The bootstrap, relying heavily upon the notion of computer simulation, provides estimates of a statistic's variance and bias. These estimates seem to be more reliable in a wide variety of settings than any previously known methods. Though extensive computation is required, the bootstrap is easy to implement on today's common microcomputers or mainframes. Many characteristics of the bootstrap technique are still unknown, but it is rapidly becoming one of the major tools for many current statisticians.

The jackknife will be discussed first in Sections 7.2 and 7.3, and then the bootstrap will be presented in Section 7.4. The exercises at the end of the chapter will serve to further illustrate the two techniques.

## 7.2. The Jackknife Algorithm

In this section, we describe the procedure involved in using the jackknife technique. In the next section, we provide some examples using the jacknife.

Suppose we wish to estimate an unknown quantity $\theta$ for some population. $\theta$ might be any of the population characteristics commonly seen in statistics: mean, median, variance, standard deviation, correlation coefficient, slope for regression, etc. Suppose that we have an estimator $\hat{\theta}$ that we might want to use, but we are unsure about its quality. $\hat{\theta}$ must come from data, so assume that $X_1, X_2, \ldots, X_n$ represent a random sample of $n$ observations from our population. These are the values used in calculating $\hat{\theta}$. But will $\hat{\theta}$ be close to the true value $\theta$?

One way to answer this question is to examine the MSE (mean square error) of $\hat{\theta}$. To do this, we must know the bias and variance of $\hat{\theta}$. We can never know them exactly (except in a simulation experiment of our creation), but the jackknife provides a way to estimate them.

Begin by calculating new statistics: Let $\hat{\theta}_{(1)}$ be the value of $\hat{\theta}$ derived from only the $(n-1)$ values $X_2, X_3, \ldots, X_n$, with $X_1$ omitted. Let $\hat{\theta}_{(2)}$ be the value of $\hat{\theta}$ derived from the $(n-1)$ values $X_1, X_3, \ldots, X_n$, with $X_2$ omitted. In general, let $\hat{\theta}_{(i)}$ be the value of $\hat{\theta}$ derived from the $(n-1)$ values $X_1, \ldots, X_{i-1}, X_{i+1}, \ldots, X_n$, with $X_i$ omitted. This can be done $n$ different times, each time leaving out exactly one of the $n$ observations $X_1, \ldots, X_n$. Thus we get $n$ new values $\hat{\theta}_{(1)}, \hat{\theta}_{(2)}, \ldots, \hat{\theta}_{(n)}$, which are called the *resampled* values of $\hat{\theta}$. These new resampled values are based on $(n-1)$ observations. Their values will probably be similar to each other, and similar to $\hat{\theta}$, but not identical. Let

$$\hat{\theta}_{(\cdot)} = \frac{\sum_{i=1}^{n} \hat{\theta}_{(i)}}{n}$$

be the average of the $n$ *resampled* statistics. By comparing $\hat{\theta}_{(\cdot)}$ to $\hat{\theta}$, we will get an idea of the bias of $\hat{\theta}$.

It can be shown that if $E(\hat{\theta}) \neq \theta$, (that is, $\hat{\theta}$ is a biased estimate of $\theta$), then an estimate of the bias is given by

$$\widehat{\text{Bias}} = (n-1)(\hat{\theta}_{(\cdot)} - \hat{\theta}) \quad . \tag{7.2.1}$$

The estimated bias is the amount by which the statistic $\hat{\theta}$ is thought to consistently overestimate or underestimate the actual value $\theta$. To correct for this consistent error of estimation, we can subtract the estimated bias to compute an adjusted statistic:

$$\begin{aligned} \tilde{\theta} &= \hat{\theta} - (n-1) \cdot (\hat{\theta}_{(\cdot)} - \hat{\theta}) \\ &= n\hat{\theta} - (n-1)\hat{\theta}_{(\cdot)} \quad . \end{aligned} \tag{7.2.2}$$

It has also been found that $\hat{\theta}_{(\cdot)}$ can be used to develop an estimate of the variance of $\hat{\theta}$. The formula is

$$\hat{\text{Var}} = \frac{n-1}{n} \sum_{i=1}^{n} [\hat{\theta}_{(i)} - \hat{\theta}_{(\cdot)}]^2 \quad . \tag{7.2.3}$$

Tukey (1958) was the first to notice the usefulness of the jackknife statistic for estimating $\text{Var}(\hat{\theta})$. His idea was to create a sample of "pseudo-values" by adjusting $\hat{\theta}$ according to each resampled value $\hat{\theta}_{(i)}$:

$$\tilde{\theta}_i = \hat{\theta} - (n-1) \cdot (\hat{\theta}_{(i)} - \hat{\theta}) \quad \text{for } i = 1, 2, \ldots, n \quad . \tag{7.2.4}$$

This new sample of pseudo-values can be averaged to arrive at the bias-adjusted estimate of $\theta$:

$$\tilde{\theta} = \frac{1}{n} \sum_{i=1}^{n} \tilde{\theta}_i \quad . \tag{7.2.5}$$

The sample variance of the pseudo-values can be used to compute the same estimate of the variance of $\hat{\theta}$ that was given in equation (7.2.3). This alternate formula is

$$\begin{aligned} \hat{\text{Var}} &= \frac{1}{n \cdot (n-1)} \sum_{i=1}^{n} (\tilde{\theta}_i - \tilde{\theta})^2 \\ &= \frac{1}{n \cdot (n-1)} \sum_{i=1}^{n} \tilde{\theta}_i^2 - \frac{1}{n-1} \tilde{\theta}^2 \quad . \end{aligned} \tag{7.2.6}$$

Notice that the pseudo-values are not really a random sample, since they are not independent of each other. But Tukey's method uses the pseudo-values as if they were a random sample, in order to calculate a sample variance and compute $\hat{\text{Var}}$.

At this point, the jackknife has given us all we need to estimate the MSE of the statistic $\hat{\theta}$:

$$
\begin{aligned}
\hat{\text{MSE}} &= (\hat{\text{Bias}})^2 + \hat{\text{Var}} \\
&= (n-1)\cdot(\hat{\theta}_{(\cdot)} - \hat{\theta})^2 + \frac{n-1}{n}\sum_{i=1}^{n}[\hat{\theta}_{(i)} - \hat{\theta}_{(\cdot)}]^2 \\
&= \frac{n-1}{n}\sum_{i=1}^{n}(\hat{\theta}_{(i)} - \hat{\theta})^2 \quad . \qquad (7.2.7)
\end{aligned}
$$

The beauty of the jackknife technique is that it allows estimation of the variance or MSE of a statistic from only *one* sample value of that statistic. Without it, we would have needed a *sample* of $\hat{\theta}$ values, calculated from repeated samples of observations, in order to estimate the variance of $\hat{\theta}$. In many cases, repeated samples are too costly to obtain, or perhaps are not even possible. Thus, any procedure, like the jackknife, that allows computation of a standard error from just one sample value, is quite attractive. Notice that $\overline{X}$ (as an estimate of the population mean $\mu$) has this property. The variance of $\overline{X}$ can be estimated from only one sample by $S^2/n$.

A 95% confidence interval for $\hat{\theta}$ can now be computed from our one sample, if we assume that $\hat{\theta}$ follows an approximate normal distribution:

$$(\hat{\theta} - 1.96\sqrt{\hat{\text{Var}}} \quad , \quad \hat{\theta} + 1.96\sqrt{\hat{\text{Var}}}\,) \quad .$$

*Remark:* A more reliable confidence interval for $\theta$ might be centered at $\tilde{\theta}$, rather than $\hat{\theta}$, since $\tilde{\theta}$ should be less biased. However, $\hat{\text{Var}}$ tends to be closer to $\text{Var}(\hat{\theta})$ rather than $\text{Var}(\tilde{\theta})$, so the issue of how to construct the confidence interval is not easily settled. We suggest using the above interval for the sake of simplicity.

## 7.3. Applying the Jackknife Procedure

To illustrate the use of the jackknife, we now present some examples where it might be useful. Further examples can be found in the exercises at the end of this chapter.

*Example 7.3.1:* To begin looking at the jackknife in action, it can be interesting to examine it in a familiar setting: estimation of a population mean $\mu$. Thus we let $\theta = \mu$. We sample $n$ observations $X_1, X_2, \ldots, X_n$. The usual estimator is $\hat{\theta} = \overline{X}$, the sample average. Of course, $\overline{X}$ is an unbiased estimate of $\theta$, and the variance of $X$ can be estimated from the sample by $S^2/n$, where $S^2$ is the sample variance. So the jackknife serves no apparent purpose in this case. If we do decide to use the jackknife, though, we have the following calculations to do:

$$\hat{\theta}_{(i)} = \frac{1}{n-1} \sum_{j \neq i} X_j$$

$$= \frac{1}{n-1} (n\overline{X} - X_i) \qquad \text{for } i = 1, 2, \ldots, n$$

$$\hat{\theta}_{(\cdot)} = \frac{1}{n} \sum_{i=1}^{n} \hat{\theta}_{(i)} = \overline{X}$$

$$\text{Bi}\hat{\text{a}}\text{s} = (n-1) \cdot (\hat{\theta}_{(\cdot)} - \hat{\theta}) = 0$$

$$\tilde{\theta} = \hat{\theta} - \text{Bi}\hat{\text{a}}\text{s} = \overline{X}$$

$$\text{V}\hat{\text{a}}\text{r} = \frac{n-1}{n} \sum_{i=1}^{n} [\hat{\theta}_{(i)} - \hat{\theta}_{(\cdot)}]^2$$

$$= \frac{n-1}{n} \sum_{i=1}^{n} \left( \frac{\overline{X} - X_i}{n-1} \right)^2$$

$$= \frac{1}{n \cdot (n-1)} \sum_{i=1}^{n} (X_i - \overline{X})^2$$

Thus the jackknife does nothing to change the usual estimates of $\mu$ and $\sigma^2/n$. (This is exactly what we would hope for; we don't want to mess up a good thing!) In fact, the pseudo-values calculated according to Eq. (7.2.4) turn out to be

$$\tilde{\theta}_i = \bar{X} - (n-1) \cdot \left( \frac{1}{n-1} \sum_{j \neq i} X_j - \bar{X} \right)$$

$$= n\bar{X} - \sum_{j \neq i} X_j$$

$$= X_i \quad \text{for} \quad i = 1, 2, \ldots, n \ .$$

This can be considered motivation for using the jackknife in other situations, since we know it gives sensible results in this case.

*Example 7.3.2:* As a further example using common statistics, let us consider estimation of the population variance $\theta = \sigma^2$, using the biased sample variance

$$S_n^2 = \frac{1}{n} \sum_{i=1}^{n} (X_i - \bar{X})^2 = \frac{1}{n} \left[ \sum_{i=1}^{n} X_i^2 - \frac{1}{n} \left( \sum_{i=1}^{n} X_i \right)^2 \right] .$$

It can be shown that $E(S_n^2) = (n-1)\sigma^2/n$. Therefore $S_n^2$ is a biased estimator of $\sigma^2$. If we jackknife $S_n^2$, we can obtain a new estimator:

$$\hat{\theta}_{(i)} = \frac{1}{n-1} \sum_{j \neq i} (X_j - \bar{X}_{(i)})^2$$

where

$$\bar{X}_{(i)} = \frac{1}{n-1} \sum_{j \neq i} X_j \ , \quad \text{for } i = 1, \ldots, n \ .$$

$$\hat{\theta}_{(\cdot)} \doteq \frac{1}{n} \sum_{i=1}^{n} \hat{\theta}_{(i)}$$

$$= \frac{1}{n} \sum_{i=1}^{n} \left[ \frac{1}{n-1} \sum_{j \neq i} (X_j - \bar{X}_{(i)})^2 \right]$$

$$= \frac{1}{n} \sum_{i=1}^{n} \left[ \frac{1}{n-1} \sum_{j \neq i} X_j^2 - \bar{X}_{(i)}^2 \right] .$$

Some involved algebraic manipulations yield

$$\hat{\theta}_{(\cdot)} = \frac{n-2}{(n-1)^2} \sum_{i=1}^{n} X_i^2 - \frac{n-2}{n(n-1)^2} \left( \sum_{i=1}^{n} X_i \right)^2 .$$

$$\begin{aligned}
\hat{\text{Bias}} &= (n-1)(\hat{\theta}_{(\cdot)} - \hat{\theta}) \\
&= (n-1) \left[ \frac{n-2}{(n-1)^2} \sum_{i=1}^{n} X_i^2 - \frac{n-2}{n(n-1)^2} \left( \sum_{i=1}^{n} X_i \right)^2 \right. \\
&\qquad \left. - \left( \frac{1}{n} \sum_{i=1}^{n} X_i^2 + \frac{1}{n^2} \left( \sum_{i=1}^{n} X_i \right)^2 \right) \right] .
\end{aligned}$$

Some more algebra shows that this is

$$\hat{\text{Bias}} = \frac{-1}{n(n-1)} \left[ \sum_{i=1}^{n} X_i^2 - \frac{1}{n} \left( \sum_{i=1}^{n} X_i \right)^2 \right] .$$

$$\begin{aligned}
\tilde{\theta} &= \hat{\theta} - \hat{\text{Bias}} \\
&= \frac{1}{n} \left[ \sum_{i=1}^{n} X_i^2 - \frac{1}{n} \left( \sum_{i=1}^{n} X_i \right)^2 \right] \\
&\qquad + \frac{1}{n(n-1)} \left[ \sum_{i=1}^{n} X_i^2 - \frac{1}{n} \left( \sum_{i=1}^{n} X_i \right)^2 \right] . \\
&= \frac{1}{n-1} \left[ \sum_{i=1}^{n} X_i^2 - \frac{1}{n} \left( \sum_{i=1}^{n} X_i \right)^2 \right] \\
&= \frac{1}{n-1} \sum_{i=1}^{n} (X_i - \overline{X})^2 . \qquad (7.3.1)
\end{aligned}$$

Thus $\tilde{\theta}$ is the unbiased form of the sample variance; i.e.

$$E(\tilde{\theta}) = \theta = \sigma^2 \quad .$$

It can also be shown that

$$\hat{\text{Var}} = \frac{n-1}{n} \sum_{i=1}^{n} (\hat{\theta}_{(i)} - \hat{\theta}_{(\cdot)})^2$$

$$= \frac{n^2}{(n-1)\cdot(n-2)^2} \left[ \sum_{i=1}^{n} (X_i - \bar{X})^4 - \left( \sum_{i=1}^{n} (X_i - \bar{X})^2 \right)^2 \right] \quad . \tag{7.3.2}$$

It should be noted that, even though the derivations of Eqs. (7.3.1) and (7.3.2) are long and difficult, one need not rely on them in the actual computation of the jackknife estimates $\tilde{\theta}$ and $\hat{\text{Var}}$. Rather, one can proceed directly according to the algorithm presented in Section 7.2. We summarize this algorithm once again:

(1) First calculate $\hat{\theta}_{(i)}$ for each $i = 1, \ldots, n$, by omitting the $i$th value of $x$ from the sample. Store them in an array for later use.

(2) Average the $\hat{\theta}_{(i)}$'s together to find $\hat{\theta}_{(\cdot)}$.

(3) Find the estimated bias: $\hat{\text{Bias}} = (n-1)\cdot(\hat{\theta}_{(\cdot)} - \hat{\theta})$.

(4) Compute the estimator adjusted for bias:

$$\tilde{\theta} = \hat{\theta} - \hat{\text{Bias}}$$

(5) Compute the estimated variance of $\hat{\theta}$:

$$\hat{\text{Var}} = \frac{n-1}{n} \sum_{i=1}^{n} (\hat{\theta}_{(i)} - \hat{\theta}_{(\cdot)})^2$$

$$= \frac{n-1}{n} \left[ \sum_{i=1}^{n} \hat{\theta}_{(i)}^2 - n\hat{\theta}_{(\cdot)}^2 \right] \quad .$$

*Example 7.3.3:* Suppose we wish to estimate $\theta = \mu^2$, where $\mu$ is the mean of a population. (This might occur, for instance, if our random variable $X$ represents the waiting time until the next occurrence is observed in a Poisson process. In this case, if $\lambda$ is the average rate of occurrence for the process, then $E(X) = 1/\lambda = \mu$ and $\text{Var}(X) = 1/\lambda^2 = \mu^2$. In order to estimate $\text{Var}(X)$, we must estimate $\mu^2$.) A natural estimator for $\theta = \mu^2$ is $\hat{\theta} = \bar{X}^2$, where $\bar{X}$ is the sample mean of $n$ observations. Note that $\bar{X}^2$ is a biased estimate of $\mu^2$:

$$\text{Var}(\bar{X}) = \sigma^2/n$$

$$E(\bar{X}^2) - \mu^2 = \sigma^2/n$$

$$E(\bar{X}^2) = \mu^2 + \sigma^2/n > \mu^2 \ .$$

Perhaps the jackknife will give us an improved estimator. We first find

$$\hat{\theta}_{(i)} = \bar{X}_{(i)}^2 = \left(\frac{1}{n-1} \sum_{j \neq i} X_j\right)^2 \quad \text{for } i = 1, 2, \ldots, n \, .$$

Then

$$\hat{\theta}_{(\cdot)} = \frac{1}{n} \sum_{i=1}^{n} \bar{X}_{(i)}^2$$

$$= \frac{1}{n(n-1)^2}\left[\sum_{i=1}^{n} X_i^2 + (n-2)\left(\sum_{i=1}^{n} X_i\right)^2\right] \ .$$

$$\widehat{\text{Bias}} = (n-1) \cdot (\hat{\theta}_{(\cdot)} - \hat{\theta})$$

$$= (n-1)\left(\frac{1}{n(n-1)^2}\left[\sum_{i=1}^{n} X_i^2 + (n-2)\left(\sum_{i=1}^{n} X_i\right)^2\right] - \frac{1}{n^2}\left(\sum_{i=1}^{n} X_i\right)^2\right)$$

$$= \frac{1}{n \cdot (n-1)} \sum_{i=1}^{n} X_i^2 - \frac{1}{n^2(n-1)}\left(\sum_{i=1}^{n} X_i\right)^2$$

$$\tilde{\theta} = \hat{\theta} - \hat{\text{Bias}}$$

$$= \frac{1}{n^2}\left(\sum_{i=1}^{n} X_i\right)^2 - \frac{1}{n\cdot(n-1)}\left(\sum_{i=1}^{n} X_i^2\right)$$

$$+ \frac{1}{n^2(n-1)}\left(\sum_{i=1}^{n} X_i\right)^2$$

$$= \frac{1}{n\cdot(n-1)}\left[\left(\sum_{i=1}^{n} X_i\right)^2 - \left(\sum_{i=1}^{n} X_i^2\right)\right]$$

$$= \frac{1}{n\cdot(n-1)} \sum_{i\neq j} X_i X_j \quad .$$

It can be shown that $E(\tilde{\theta}) = \mu^2$ (see Ex. 7.5), so that $\tilde{\theta}$ is an unbiased estimate of $\mu^2$. This is not necessarily true of all jackknife estimators, but is fairly common, as we have seen in these examples.

Finally, the estimate of $\text{Var}(\bar{X}^2)$ can be found directly according to Eq. (7.2.3).

## 7.4. The Bootstrap Algorithm

We now look at another method of gauging the accuracy of an estimate, $\hat{\theta}$. The bootstrap is a new technique, having been introduced by Brad Efron of Stanford University, as recently as 1979. The procedure is described in the following.

Collect a random sample of observations $x_1, x_2, \ldots, x_n$ for calculating the usual estimate $\hat{\theta}$. We use small $x$'s here, rather than capitals, to emphasize the fact that the remainder of the algorithm is based upon having already observed these $n$ values. Now do a computer simulation based on the $n$ values from the sample. The simulation should proceed by generating repeated samples of $n$ values from an artificial discrete distribution. This artificial distribution comes from the original sample: give each value in the original sample a probability of $1/n$. Thus, we generate repeated samples of $n$ values from a dicrete distribution with

$$P(X = x_i) = 1/n \quad , \quad \text{for } i = 1, 2, \ldots, n \quad .$$

Naturally the repeated samples will not be exactly the same as the original set of $n$ values. The repeated samples are generated randomly, and duplicates of a value within a sample are likely to occur. However, the repeated samples should have some characteristics similar to the original $n$ values. In particular, an estimate of $\theta$, say $\hat{\theta}*$, can be calculated from each new sample. Suppose we generate $B$ new samples. Then we also generate $B$ new estimates of $\theta$. Let us denote these by $\hat{\theta}_1*, \hat{\theta}_2*, \ldots, \hat{\theta}_B*$. These estimates can be thought of as an approximation to the sampling distribution of $\hat{\theta}$. The mean of the $\hat{\theta}*$ estimates should be approximately equal to the original $\hat{\theta}$, as well as to the true value of $\theta$. The variance of the $B$ values of the $\hat{\theta}*$ estimates should be approximately equal to the variance of the sampling distribution of $\hat{\theta}$, i.e. $\mathrm{Var}(\hat{\theta})$. Thus we have the estimate of variance that was desired:

$$\hat{\mathrm{Var}} = \frac{1}{B} \sum_{i=1}^{B} (\hat{\theta}_i* - \bar{\hat{\theta}}*)^2 \quad ,$$

where $\bar{\hat{\theta}}* = (1/B) \sum_{i=1}^{B} \hat{\theta}_i*$ is the average of the repeated sample estimates of $\theta$. All we have done is to compute the sample variance of the $\hat{\theta}*$'s. (Note: The issue of whether to divide by $B$ or $B-1$ in calculating this variance will be ignored. $B$ should be large enough so that the resulting difference in the estimated variance will be negligible.)

An estimate of the bias of $\hat{\theta}$ is available from

$$\mathrm{Bias} = \bar{\hat{\theta}}* - \hat{\theta} \quad .$$

An adjusted bootstrap estimator of $\theta$ may then be found from

$$\tilde{\theta}* = \hat{\theta} - \hat{\mathrm{Bias}} \quad .$$

We interpret this to mean that if $\bar{\hat{\theta}}*$ is larger than the original $\hat{\theta}$, this is an indication that the estimate $\hat{\theta}$ tends to be larger than the true value of $\theta$. Similarly, if $\bar{\hat{\theta}}*$ is smaller than the original $\hat{\theta}$, this is an indication that the estimator $\hat{\theta}$ tends to be smaller than the true value of $\theta$.

This bootstrap procedure can be implemented on any statistic. Computer simulation is a necessary part of the procedure. Such a procedure could not have been considered years ago, before the computer age. But now, repeated samples of $n$ values can be generated with such speed, that the procedure is far from laborious. In fact, this method is usually far simpler than attempting to work out the exact bias and variance of $\hat{\theta}$ mathematically. Why bother with the difficult mathematics, when the computer can do the job for us — both faster and cheaper!

Efron has demonstrated that the number of repeated samples ($B$) does not need to be extremely large. In most cases, $B = 100$ or $B = 200$ samples provide enough estimates $\hat{\theta}_i^*$, to make the bootstrap useful. More samples than this usually provide little additional information.

The bootstrap algorithm can be applied to the same statistics for which the jackknife has been used earlier in this chapter. We begin with a simple example, where $\theta = \mu$ represents an unknown population mean to be estimated. Let $\hat{\theta} = \overline{X}$ represent the sample mean found from $n = 10$ observations. In particular, suppose we have observed these 10 values:

2.30 2.57 2.98 3.12 3.21 3.23 3.39 3.45 3.70 3.92

(The values have been numerically ordered for simplicity.) We can easily find $\hat{\theta} = \overline{X} = 3.187$. Now in reality, we know $\overline{X}$ is a good estimate of $\mu$. In fact, we know it is unbiased and has a variance approximated by the sample variance $S^2$ divided by $n$. In this case, we have

$$S^2 = \frac{1}{9} \sum_{i=1}^{10} (x_i - 3.187)^2 = .236 \ ,$$

so that $S^2/n = .0236$. However, we can still use the bootstrap in order to compare the results.

To find bootstrap estimates of the bias and variance of $\overline{X}$, we must generate repeated samples using the 10 values observed above. One way to generate the samples is to store the values in an array:

$$X(1) = 2.30 \quad X(2) = 2.57 \quad \ldots \quad X(10) = 3.92 \ .$$

Now to generate random values from a discrete uniform distribution with the above values, we can simply generate random values for the array index from 1 to 10, and output the associated values of the array $X$. For example, from Chapter 1 we know that

$$I = \text{INT}(10^* U + 1) \ ,$$

where $U$ is a uniform random number between 0 and 1, produces a random integer value from 1 to 10.

Using this idea, we present the following algorithm for the bootstrap:

1) Store the $n$ values in array $X$: $X(1) \ldots X(n)$.
   (In this case, $n = 10$.)
2) For $k = 1$ to $B$. ($B$ repeated samples)
3) For $j = 1$ to $n$. ($n$ values in each sample)

4) Let $I = \text{INT}(10*U + 1)$ where $U$ is the next random number.
5) Output $Y(j) = X(I)$ the $j$th value of the repeated sample stored in an array $Y$: $Y(1) \ldots Y(n)$.
6) Next $j$.
7) Compute $\hat{\theta}_k{}^* = \dfrac{1}{n} \sum_{j=1}^{n} Y(j)$ and store $\hat{\theta}_k{}^*$ in an array for later use.

   (In this case, $\hat{\theta}_k{}^*$ is the sample mean from the $k$th repeated sample.)
8) Next $k$.
9) Compute $\bar{\hat{\theta}}{}^* = \dfrac{1}{B} \sum_{k=1}^{B} \hat{\theta}_k{}^*$, the average of the repeated sample estimates.
10) Compute $\widehat{\text{Bias}} = \bar{\hat{\theta}}{}^* - \hat{\theta}$. (In this case, $\theta = 3.187$, the average of the original 10 values. We expect Bias to be near 0, since we know our statistic $\hat{\theta} = \bar{X}$ is really unbiased.)
11) Compute $\widehat{\text{Var}} = \dfrac{1}{B} \sum_{k=1}^{B} (\hat{\theta}_k{}^* - \bar{\hat{\theta}}{}^*)^2$

    or $\qquad \widehat{\text{Var}} = \dfrac{1}{B} \sum_{k=1}^{B} (\hat{\theta}_k{}^*)^2 - (\bar{\hat{\theta}}{}^*)^2$

    (We expect $\widehat{\text{Var}}$ to be near $S^2/n = 0.0236$.)

We know what to expect from the bootstrap when using $\hat{\theta} = \bar{X}$ as our estimator, since its properties are well known. In other cases, however, the bias and variance of an estimate $\hat{\theta}$ may not be known. For example, suppose we wish to estimate the population variance $\theta = \sigma^2$. We can use $\hat{\theta} = S^2$, where $S^2$ is the usual sample variance, dividing by $n - 1$ if we wish it to be unbiased. But what about the variance of the estimate $S^2$? We have seen how to estimate it using the jackknife (Example 7.3.2), but we might instead use the bootstrap to approximate it. The only change that needs to be made in the algorithm presented above, is to calculate

$$\bar{Y} = \frac{1}{n} \sum_{j=1}^{n} Y(j)$$

and $\hat{\theta}_k{}^* = \dfrac{1}{n-1} \sum_{j=1}^{n} (Y(j) - \bar{Y})^2$ .

Then this $\hat{\theta}_k{}^*$ is stored in the appropriate array, and later used in the calculation of $\bar{\hat{\theta}}{}^*$ (the average of the repeated sample estimates of $\theta$) and in the calculation of $\hat{\text{Var}}$, the estimated variance of $\hat{\theta}$.

Thus, the bootstrap provides an easy way to estimate the bias and variance of an estimate $\hat{\theta}$. The bootstrap technique can be used in many of the cases for which a jackknife can be used. The general goals of the two procedures are roughly the same. The exercises at the end of the chapter provide more examples of the use of the bootstrap.

Which procedure is better, the jackknife or the bootstrap? Efron, in 1983, has given evidence that the bootstrap estimate of variance tends to be better than the jackknife estimate of variance, in that it more closely approximates the true variance of $\hat{\theta}$. This would indicate that the bootstrap procedure is preferable to the jackknife. In addition, the bootstrap algorithm is easily carried out with a computer program that supplies the repeated bootstrap samples.

The bootstrap offers still another bit of flexibility over the jackknife. In forming a confidence interval for the true value of a parameter $\theta$, the best option using the jackknife procedure is to hope that the sampling distribution of the original estimator $\hat{\theta}$ is approximately normal. Then a $100(1-\alpha)\%$ confidence interval can be formed using $\hat{\text{Var}}$ found from the jackknife algorithm:

$$(\hat{\theta} - Z_{1-\alpha/2}\sqrt{\hat{\text{Var}}} \quad , \quad \hat{\theta} + Z_{1-\alpha/2}\sqrt{\hat{\text{Var}}}) \quad . \tag{7.4.1}$$

However, the normality of $\hat{\theta}$ is often not known, particularly for a small sample size $n$.

A confidence interval from the bootstrap algorithm need not necessarily rely on the assumption of a normal distribution for $\hat{\theta}$. (An interval may be formed using (7.4.1) and using $\hat{\text{Var}}$ from the bootstrap algorithm.) An alternative procedure is to construct a bootstrap histogram from the repeated bootstrap sample estimates $\hat{\theta}_k{}^*$, for $k = 1, 2, \ldots, B$. The $100(\alpha/2)$ and $100(1-\alpha/2)$ percentiles of this histogram would provide a $100(1-\alpha)\%$ confidence interval for $\theta$.

For example, suppose $B = 200$ repeated samples are taken. Then 200 bootstrap estimates $\hat{\theta}_k{}^*$ can be calculated and stored in an array $B$: $B(1)$, $B(2), \ldots, B(200)$. To find percentiles in these 200 values, the estimates should be sorted numerically in the array, so that

$B(1)$ contains the smallest value,

$B(2)$ contains the second smallest value etc.,

$B(200)$ contains the largest value.

Then a 95% confidence interval for $\theta$ would use the 2.5 percentile ($B(5)$) and the 97.5 percentile ($B(195)$). The confidence interval would then be ($B(5)$, $B(195)$). The sorting of the values can be done quickly and easily using any of today's modern sorting techniques (for example, the quicksort in the next chapter).

**Exercise 7** (*All programming exercises except 7.5*)

7.1 The Pearson product moment correlation coefficient is used to measure the amount of linear association between two variables, $X$ and $Y$. If $(X_1, Y_1)$, $(X_2, Y_2)$, . . . , $(X_n, Y_n)$ represents a random sample of $n$ bivariate observations, then the Pearson correlation is found by (5.6.4)

$$\hat{p} = \frac{\sum_{i=1}^{n} (X_i - \overline{X}) \cdot (Y_i - \overline{Y})}{\sqrt{\sum_{i=1}^{n} (X_i - X)^2 \sum_{i=1}^{n} (Y_i - Y)^2}} .$$

(a) Write a computer program that generates 25 pairs $(X_i, Y_i)$, where each $X_i$ has a standard normal distribution ($\mu = 0$, $\sigma^2 = 1$) and $Y_i = X_i + \epsilon_i$, where $\epsilon_i$ also has a standard normal distribution, independent of $X_i$. Use the 25 $(X_i, Y_i)$ pairs to compute the Pearson correlation coefficient.

(b) Compute a jackknife correlation coefficient, by getting an estimate of the bias in $\hat{p}$ from part (a) and subtracting that bias.

(c) Compute an estimated variance for the correlation $\hat{p}$, using the jackknife. Use that variance to construct a 95% confidence interval for the population correlation coefficient.

(d) Use the bootstrap technique on the 25 $(X_i, Y_i)$ values that you have generated, by storing them in a 25 × 2 array and giving each $(X_i, Y_i)$ pair a probability of $1/25 = 0.04$. Randomly resample the $n = 25$ pairs until you have compiled $B = 100$ samples with 25 pairs in each sample. From each sample, compute the correlation coefficient. Store these 100 correlation coefficients in an array, and use them to find an estimate of the variance. How does it compare to the jackknife variance from part (c)?

7.2 Suppose we observe customers arriving at a queue according to what we think is a Poisson process; i.e. the inter-arrival times are independent and identically distributed exponential random variables defined by (2.4.4). We can observe inter-arrival times $X_1, X_2, \ldots, X_n$, which would follow an exponential distribution under the assumption of a Poisson process for the arrivals. If the expected inter-arrival time is $E(X) = \mu$, then the average rate of arrival is given by $\lambda = 1/\mu$.

(a) Generate 20 inter-arrival times by generating 20 random numbers from an exponential distribution with mean = 1. Compute the estimated arrival rate $\lambda = 1/\overline{X}$, where $\overline{X}$ is the sample average of the 20 generated inter-arrival times.

(b) Compute the estimated bias of $\hat{\lambda}$ by using the jackknife technique. Then compute a new estimated arrival rate, by adjusting $\hat{\lambda}$ for its bias.

(c) Use the jackknife to estimate the variance of $\hat{\lambda}$, and then form a 95% confidence interval for the true $\lambda$. (Note: In the simulation from part (a), the true value is $\lambda = 1$.)

(d) Use the bootstrap technique on the 20 values generated in part (a). Derive a total of $B = 125$ resampled bootstrap estimators, using $n = 20$ values in each sample. Find the estimated bias and variance from these values.

7.3 We know that the sample variance

$$S^2 = \frac{1}{n-1} \sum_{i=1}^{n} (X_i - \overline{X})^2$$

is an unbiased estimate of the population variance $\sigma^2$. It is not true, however, that the sample standard deviation $S = \sqrt{S^2}$ is an unbiased estimate of $\sigma$. In fact, it can be shown that $E(S) < \sigma$, when the $X$'s follow a normal distribution.

(a) Generate a random sample of 5 values from a standard normal distribution. Calculate the sample standard deviation $S$.

(b) Estimate the bias in $S$ by jackknifing it. What would be the adjusted value for $S$?

(c) Estimate the variance of $S$ from the jackknife. Form a 95% confidence interval for the true $\sigma$.

(d) Which is the better estimate of $\sigma$, $S$ or the jackknifed value found in part (b)? Answer this question by doing a simulation. Generate 1000 samples of size 5 from the standard normal distribution. For each sample, calculate the usual sample standard deviation $S$ and jackknifed bias-adjusted estimate of $\sigma$. Compute the absolute error and squared error of each. (Remember that the true value is $\sigma = 1$.) Compare the two methods of estimating $\sigma$ by comparing the average absolute error and average squared error over the 1000 samples.

7.4 Let us consider a binomial experiment. We can observe $n$ Bernoulli trials, where $p$ is the unknown probability of success on any one trial. The estimate is $\hat{p} = X/n$, where $X$ is the number of successes observed from the $n$ trials. To form an approximate confidence interval for $p$, an estimate of $\text{Var}(\hat{p}) = p(1-p)/n$ is required. Let us estimate $\theta = p(1-p)$ in order to know $\text{Var}(\hat{p})$. A natural estimator would use the sample $\hat{p}$ in place of the unknown $p$; i.e. $\hat{\theta} = \hat{p}(1-\hat{p})$. In order to compute the jackknifed version of this estimator, we must consider the $X$ successes and

$n - X$ failures as comprising our sample of $n$ observations. For instance, suppose we observe 4 successes and 7 failures from a sample of 11 Bernoulli trials. Use this information to answer the following question. For simplicity, you may assume that $X_1$, $X_2$, $X_3$, and $X_4$ represent successes, and $X_5, X_6, X_7, X_8, X_9, X_{10}$, and $X_{11}$ represent failures.

(a) Find $\hat{\theta}$, the usual estimate of $p(1-p)$.

(b) What would be the values of the 11 resampled estimators, $\hat{\theta}_{(i)}$, obtained by leaving out each of the 11 Bernoulli trials, in turn? Note that the 11 resampled estimators separate themselves into two groups, according to whether a success or failure has been omitted.

(c) Find the average $\hat{\theta}_{(\cdot)}$ and then compute the estimated bias. Compute the bias-adjusted estimate $\tilde{\theta}$. (It can be shown that, in general, the bias-adjusted estimate of $\theta$ will be $n\hat{p}(1-\hat{p})/(n-1) = n\hat{\theta}/(n-1)$.

(d) Find the estimated variance of $\hat{\theta}$, according to Eq. (7.2.3). Assuming a normal distribution of $\hat{\theta}$, find a 95% confidence interval for $\theta = p(1-p)$.

(e) Use the bootstrap on the given 11 Bernoulli trials. Generate 200 samples with 11 values (Bernoulli trials) in each sample. Find the bootstrap estimator of $\theta = p(1-p)$, adjusted for bias.

(f) Find the bootstrap estimate of variance for $\hat{\theta}$. Order the 200 bootstrap estimators to find a 95% confidence interval for $\theta$, using the 2.5 percentile and the 97.5 percentile of the ordered values. Compare your answers here with those in parts (c) and (d).

7.5 Show that if $\theta = \mu^2$ and $\tilde{\theta} = \dfrac{1}{n \cdot (n-1)} \sum_{i \neq j} X_i X_j$, then $\tilde{\theta}$ is an unbiased estimate of $\mu^2$. (See Example 7.3.3.) Hint: $E(X_i X_j) = E(X_i) \cdot E(X_j)$ for any $i \neq j$, since $X_1, \ldots, X_n$ is a random sample of independent values.

7.6 Consider the estimation of $\theta = \mu^2$, presented in Example 7.3.3.

(a) Generate a sample of 10 inter-arrival times by generating 10 random numbers from an exponential distribution with mean = 1. Compute the usual estimate $\theta = \overline{X}^2$ from your sample.

(b) Compute the bias-adjusted jackknife estimator, $\tilde{\theta}$.

(c) Compute the jackknife estimate for Var($\hat{\theta}$).

(d) Using the 10 sample values generated in part (a), create a bootstrap sample of estimates. Use $B = 100$ repeated samples. Find the bootstrap estimate of $\mu^2$, by averaging the 100 bootstrap sample estimates together. Estimate the bias in this estimate and adjust the bootstrap estimate for this bias.

(e) Find the bootstrap estimate for Var($\hat{\theta}$).

(f) What is the actual value of Var($\hat{\theta}$)?

Answer this with a simulation of 100 samples of $n = 10$ random observations from the exponential distribution with mean = 1. From each sample, compute $\hat{\theta} = \overline{X}^2$. Then find the variance of these 100 sample values of $\hat{\theta}$. (Note the similarity of this procedure to the bootstrap procedure.) Based on the results of this simulation, which procedure (jackknife or bootstrap) gave the closest value to Var($\hat{\theta}$)?

Chapter 8

# STATISTICAL APPLICATIONS TO COMPUTER SCIENCE

## 8.1. Probabilistic Algorithms

Statistics and probability have become more and more useful in doing basic computer science research in areas such as software reliability, probabilistic algorithms, time sharing, and switching circuits. In order to keep the background knowledge as limited as possible, only the first two topics will be included here. Probabilistic algorithms are the discussion of this and the next section.

Most computing algorithms are deterministic in the sense that if the input is given, the programmer can predict each of the computational steps, if he so wishes. An algorithm that uses random numbers to determine its computational paths is generally referred as a probabilistic algorithm. Strictly speaking, if congruential random number generators are used for random number generation (see Section 1.3), the computational steps are still predictable by the structure of the random number generator. However, from a practical point of view, nobody cares to predict the random numbers from a random generator. Thus, one may consider that random numbers are unpredictable, and the program that uses them is nondeterministic.

The basic idea of probabilistic algorithms can be demonstrated by a simple example. It is the same idea as in practising statistics. Suppose there are $N = 100{,}000$ income tax returns in the data file and we wish to find the highest $n = 1{,}000$ incomes. An intuitive way to do this is first to find the highest income by comparing the incomes one by one, and always keeping the higher one. After $N - 1$ comparisons, the highest income is found. The second highest income can be found the same way by first skipping the already-found highest income and then doing $N - 2$ comparisons. If this process is continued 1000 times, the highest 1000 incomes are found. The total number of comparisons used to finish this procedure is

$$(N-1)+(N-2)+,\ldots+(N-n) = nN-n(n+1)/2$$
$$\approx nN$$
$$= 10^8 \; .$$

However, if we know that at least 1000 incomes are more than \$200,000, then we may eliminate all the data that are less than \$200,000 by a single comparison for each item. Suppose in doing this, we find that there are 2000 incomes that are more than \$200,000. What we then need to do is find the largest 1000 among a data file of 2000. By the same method just described we can finish this task by

$$1000 \times 2000 - 1000 \times 1001/2 \approx 1.5 \times 10^6$$

comparisons. Adding the first 100,000 comparisons in the screening process, the total number of comparisons used here is found to be approximately $1.6 \times 10^6$. This is a tremendous saving over the intuitive procedure's $10^8$ comparisons.

In practice, the threshold \$200,000 may be unknown. But we can take a sample to estimate it. From the above computation we see that if the upper 1 percentile is conservatively estimated by the upper 2 percentile, the time saved is still worthwhile. However, there is a risk that the threshold estimated percentile is too high to contain the top 1000 incomes. In this case, the process has to be repeated and the computing time saved may not be so great and may be even worse than a deterministic algorithm. Whether we can take this risk depends on the situation. Usually the merit of a probabilistic algorithm is a trade-off between its average performance time and the worst case performance time. In the next section, we will discuss a very famous and commonly used probabilistic algorithm, the quicksort.

## 8.2. The Quicksort

Sorting is a basic operation in computer science and statistics. Its applications are in database management, searching, and nonparametric statistics. (See Section 4.7). Imagine how difficult it is to find a student's record by his social security number when the records are not filed according to the order of their social security numbers. Similar difficulty appears in finding one's name when the file is not in alphabetical order. Although most stored data are often sorted by one variable, new sorting is necessary when the data are to be studied by other variables. For example, tax returns may be sorted under the order of the social security number, but we may wish to find incomes by occupation. Thus, an efficient sorting algorithm is constantly required. Almost all data management systems and mathematical and statistical packages have an efficient sorting algorithm.

Suppose there are $n$ pieces of data to be sorted. To illustrate our first sorting algorithm, we use the following example. Let the data be

10, 9, 7, 4, 12, 14, 6.

The first step is to find the smallest number and its position by comparing the numbers consecutively from the beginning to the last data. By $n - 1 = 6$ comparisons, we find that 4 is the smallest. Then 4 is exchanged with the left most number, 10. Thus, after the first round the data set becomes

4, 9, 7, 10, 12, 14, 6.

Then the second round of comparisons will start at the second number from the left. The smallest number to the right of the first number will be found and exchanged with the second number. Thus the data file now looks like

4, 6, 7, 10, 12, 14, 9.

This process will continue and the data arrays after each round of comparisons are consecutively,

4, 6, 7, 10, 12, 14, 9

4, 6, 7, 10, 12, 14, 9

4, 6, 7, 9, 12, 14, 10

4, 6, 7, 9, 10, 14, 12

4, 6, 7, 9, 10, 12, 14 .

The total computing time for accomplishing this process can be measured by the total number of comparisons and the total number of exchanges. The total number of comparisons is

$$T_c = (n-1) + (n-2) + \ldots + 3 + 2 + 1 = n(n-1)/2$$

$$\approx n^2/2 \ . \tag{8.2.1}$$

The total number of exchanges is no more than $n$, because there are situations where no exchange is necessary, as in the example. For large $n$, $n^2/2 \gg n$. Hence we may use $T_c$ as the measure of the computing time for this procedure, which is usually referred to as the exchange sort.

Now suppose that we know the median of this data set. Then the data can be easily partitioned into two groups: one with elements larger than or equal to the median, and one with elements smaller than it. Roughly speaking, each group contains $n/2$ elements. If the exchange sort is used in both groups, then the combined set is the completely sorted data file. It requires

$$2[(n/2)^2/2] \approx n^2/4$$

number of comparisons. Adding the $n-1$ number of comparisons in data partitioning, the total number of comparisons for sorting when the median is known is

$$(n-1) + n^2/2 \quad ,$$

which is approximately half of the previous sorting time. Of course, the computing time can be further reduced if the medians of the two groups are known. New ordered groups can be formed and computing time can again be reduced if the medians of those groups are known. But in practice, the medians are unknown. Hoare (1962) suggests that we estimate these medians each by a sample of size 1. He names this method quicksort which turns out to be one of the quickest and most commonly used sorting algorithms.

The quicksort procedure is very easy to describe. First, a piece of data is randomly picked and it is used as a fence for partitioning the whole data file. All the data that are larger than the fence are put into one partition and all the rest into another one. Then in each partition the same procedure of randomly picking a fence and partitioning is repeated until the partition contains only 0 or 1 element. When all the partitions are combined, a completely ordered file is formed.

To investigate the complexity of this procedure, let the data size be $n$ and the average number of comparisons needed to finish the sorting be $S(n)$. Suppose the first fence partitions the data into two groups with sizes $m$ and $n-1-m$ respectively. Then the average time to sort the two groups is $S(m) + S(n-m-1)$. Because the fence is randomly chosen, $m$ has equal chances to be $0, 1, \ldots, n-1$. Thus,

$$S(n) = n - 1 + \frac{1}{n} \sum_{m=0}^{n-1} \{S(m) + S(n-m-1)\} \; ,$$

where the first term $n-1$ is the number of comparisons used to partition the data. Obviously,

$$S(n) = n - 1 + \frac{2}{n} \sum_{m=0}^{n-1} S(m)$$

$$= n - 1 + \frac{2}{n} S(n-1) + \frac{2}{n} \sum_{m=0}^{n-2} S(m) \ . \qquad (8.2.2)$$

Since the first term implies $\sum_{m=0}^{n-1} S(m) = n(S(n)-(n-1))/2$, we have $\sum_{m=0}^{n-2} S(m) = (n-1)(S(n-1)-(n-2))/2$. Substituting it to the second equation of (8.2.2), it becomes

$$\frac{S(n)}{n+1} = \frac{S(n-1)}{n} + \frac{4}{n+1} - \frac{2}{n} \ .$$

Applying the above equation recursively, we get

$$\frac{S(n)}{n+1} = \frac{S(n-2)}{n-1} + 4\left(\frac{1}{n+1} + \frac{1}{n}\right) - 2\left(\frac{1}{n} + \frac{1}{n-1}\right)$$

$$\vdots$$

$$= \frac{S(1)}{2} + 4\left(\frac{1}{3} + \frac{1}{4} + \ldots + \frac{1}{n+1}\right)$$

$$- 2\left(\frac{1}{2} + \frac{1}{3} + \ldots + \frac{1}{n}\right) .$$

By $S(1) = 0$ and the well-known formula

$$1 + \frac{1}{2} + \frac{1}{3} + \ldots + \frac{1}{n} = \Gamma + \ln(n) \ ,$$

where $\Gamma = 0.577$ is the Euler constant, we have

$$S(n)/(n+1) \approx 4(\ln(n+1) + \Gamma - 3/2) - 2(\ln(n) + \Gamma - 1)$$

or,

$$S(n) \approx 2n\ln(n) \quad . \tag{8.2.3}$$

The difference between (8.2.1) and (8.2.3) can be tremendous for large $n$. For example, if $n = 10^6$, $2n\ln(n) = 27.6 \times 10^6$, and $n^2/2 = 5 \times 10^{11}$. If your computer can do one million ($10^6$) comparisons per second, the time difference between the two sorting algorithms is 27.6 seconds versus 5 days and 19 hours.

The real time spent in quicksort is a little more than merely comparisons. Random sampling and record keeping for the partitions take time. However, for large $n$, these times are negiligible compared to the comparisons. Table 8.2.1 gives some actual sorting time using a microcomputer. The time saved by the quicksort for moderate $n = 500$ is substantial. The last two columns give the ratio between actual time and the number of comparisons predicted by theory. They are quite consistent, though the quicksort requires more overhead.

Table 8.2.1

| $n$ | QT | ET | $QT/(2n\ln(n))$ | $ET/(n^2/2)$ |
|---|---|---|---|---|
| 100 | 6 | 25 | 0.0065 | 0.005 |
| 500 | 38 | 620 | 0.0061 | 0.005 |

QT: sorting time in seconds by the quicksort
ET: sorting time in seconds by the exchange sort

The worst case in quicksort happens when the fences picked are all very far away from the median, namely, either the smallest or the largest element is picked all the times. Under this circumstance,

$$S(n) = n - 1 + S(n-1)$$

$$= n - 1 + n - 2 + S(n-2)$$

$$\vdots$$

$$= n - 1 + n - 2 + \ldots + 2 + 1$$

which is the same as the number of comparisons required by the exchange sort. Of course, the chance of getting into this situation is extremely small (see Ex. 8.3).

A natural question arises whether one can find some other sorting methods that are much faster than quicksort. This goes to the problem of finding the minimum number of comparisons required for sorting an arbitrary data file of size $n$. Problems of this type belong to information theory, but the answer for sorting is not difficult. Let us consider binary numbers, i.e. each digit of a number takes the value 0 or 1. It is apparent that a one-digit binary number requires one comparison to recognize and a two-digit number requires two comparisons (digitwise) to recognize. Similarly a $k$ digit binary number requires $k$ comparisons to recognize. To sort a file of $n$ pieces of data is equivalent to finding out the original order of the data file which has, in general, $n!$ variations. To represent $n!$ pieces of information requires at least $\log_2(n!)$ digits. Thus, at least $\log_2(n!)$ comparisons are necessary to sort $n$ pieces of data. By the well-known Sterling's formula $n! \approx \sqrt{(2\pi n)}e^{-n} n^n$ , we have

$$C_{\min} \approx n \log_2(n) - n \approx 1.386 n \ln(n).$$

Comparing with the expected number of comparisons $2n \ln(n)$ by the quicksort, improvement is still possible but not substantial. New methods (see Fraser & McKeller (1970)) have been invented to do quicker sorting, but due to the additional programming effort, they are not often used.

## 8.3. Software Testing and Reliability

Unlike hardware, software will not wear out. The reliability of software does not decrease with time, although time might make it obsolete. Thus, the reliability of a piece of software is determined at the time it is released. In this section we will discuss how to assess the initial reliability of software.

Basically there are two categories of methods that are used for software reliability testing. Their application depends on the difficulty in checking the correctness of the output. For example, in a telephone exchange software system, the correctness of one output is very easy to check. If the call number does not reach the right telephone, it is wrong, otherwise it is correct. For this type of system, we can test a large amount of input without too much effort.

On the other hand, a statistical package is not that easy to verify. The output validity has to be checked by time-consuming hand computation. Moreover, the input data have a tremendous amount of variation that makes it difficult to find typical testing samples. Hence, the programmer usually sends free copies to a few potential users and asks them to report to him their difficulties. This is the second type of testing.

For the first kind of testing, the software will be released only when it has been tested a long time without failure. Suppose a program has been tested $n$ times without failure, how should we state its reliability? This can be answered by the concept of a confidence interval. Let $p$ be the true failure rate of the program and $p_0$ be its $(1-\alpha)$ confidence upper bound. Then obviously,

(i) if $p \leqslant p_0$, our bound is correct and we have nothing to worry about;

(ii) if $p > p_0$, then our bound is incorrect, but the probability of this situation is $(1-p)^n$, i.e., no failure in $n$ trials. We wish

$$(1-p)^n \leqslant \alpha \quad ,$$

or $p \geqslant 1-\alpha^{1/n}$. Thus, we are safe to use $p_0 = 1-\alpha^{1/n}$ as the upper bound for $p$. By using $(1-p_0)^n = \alpha$ for $1-\alpha^{1/n} = p_0$, we have

$$\exp(-np_0) \approx \alpha \quad , \text{ or } \quad p_0 = -\ln(\alpha)/n \quad . \tag{8.3.1}$$

*Example 8.3.1:* An airline ticket purchase system has been tested 1000 times without failure. Find a 95% upper bound for the failure rate.

*Solution:* By (8.3.1) $p_0 = -\ln(0.05)/1000 = 0.003$. Or by the original formula $p_0 = 1-0.05^{1/1000} = 0.003$.

In the second case, suppose the program is sent to two users. After their reports, the programmer found $r$ errors in which $n_1$ errors were due to the report of user 1, $n_2$ errors due to user 2, and there are $n_{12}$ errors in common. Then obviously

$$r = n_1 + n_2 - n_{12} \quad .$$

What does this tell us? Can we use the above information to infer the remaining errors in the program? We may reason that the problem is the same as estimating the number of red balls (errors) in a bag of many balls. Suppose the red balls are equally likely to be picked (Note: this assumption is reasonable in program testing only if all the "easier" ones have been removed by the programmer and the rest of the errors are difficult and approximately equally hard to find.) Let $e$ be the proportion of red balls that tester 1 has found, which represents his effort and ability to find red balls. Then if the total number of red balls is $N$, we should have

$$e \approx n_1 / N$$

By the same reason, $e \approx n_{12} / n_2$. Equating the two $e$'s, we get

$$N \approx n_1 n_2 / n_{12} \quad .$$

The same estimate for $N$ appears from the point of view of tester 2. Thus we may use

$$\hat{N} = n_1 n_2 / n_{12} \tag{8.3.2}$$

as the point estimate of the total errors. Then the estimate of the unfound errors is

$$\hat{U} = \hat{N} - r = (n_1 - n_{12})(n_2 - n_{12})/n_{12} \quad . \tag{8.3.3}$$

To find a confidence interval for $N$, we use the fact that the more red balls in the urn, the harder for two persons to pick up the same balls. Thus we would use $N_0$ as an upper bound for the real number of red balls, when the likelihood of finding $n_{12}$ balls in common when $n_1$ and $n_2$ balls are picked by two persons is very small, i.e., smaller than some number $\alpha$. The probability of getting more than $n_{12}$ balls from the point of view of tester 1 is

$$p = \sum_{j=n_{12}}^{n_1} \binom{n_2}{j} \binom{N-n_2}{n_1 - j} \bigg/ \binom{N}{n_1} \quad . \tag{8.3.4a}$$

It can be shown that this probability is the same from the point of view of tester 2, i.e.

$$p = \sum_{j=n_{12}}^{n_2} \binom{n_1}{j} \binom{N-n_1}{n_2 - j} \bigg/ \binom{N}{n_2} \quad . \tag{8.3.4b}$$

The smallest $N$, say, $N_0$, that makes $p$ smaller than $\alpha$ is the $(1-\alpha)$ upper confidence bound for $N$. The $p$'s in (8.3.4) can be easily determined by a computer. Here we will show two simple examples.

*Example 8.3.2:* Find the 95% upper bound for $N$, when $n_1 = n_{12} = 1$, and $n_2 = 2$.

*Solution:* The point estimate of $N$ is $n_1 n_2 / n_{12} = 2$. But since the errors found by the two testers are very small, we can't exclude the possibility that neither of them have done a good job. Hence we may expect a much higher confidence bound for $N$. By (8.3.4),

$$p = \binom{2}{1}\binom{n-2}{0}\Big/\binom{N}{1} = 2/N \ .$$

For $p = \alpha = 0.05, N = 40$. Thus the upper 95% confidence bound for $N$ is 40.

*Example 8.3.3:* If $n_1 = n_2 = 100$, and $n_{12} = 99$, estimate the unfound errors with 95% and 99% confidence bounds for it.

*Solution:* By (8.3.3), the point estimate of the unfound errors is

$$(100-99)(100-99)/99 = .01 \ .$$

It can be seen that (8.3.4) becomes

$$p = \{100(N-100)+1\}\binom{N}{100} \ .$$

Since 101 errors have already been found, we start with $N = 101$. It is easy to find

| $N$ | $p$ |
|---|---|
| 101 | 1.00 |
| 102 | 0.04 |
| 103 | 0.0017 |

Thus, the 95% confidence upper bound for the unfound errors is $102 - 101 = 1$ and the 99% bound is $103 - 101 = 2$.

As we have noted before, the present method is based on the assumption that all the errors have the same probability of being found. If the programmer feels that this is not a reasonable assumption and there are harder errors, then the present formulas give liberal estimates of the remaining errors. If the number is large, it is probably too early to release the program.

Of course, the programmer can also use this method to evaluate the remaining errors by himself. However, it is usually difficult for one person to generate two "independent" test sets. One method suggested in the literature (see Myers (1976)) is to insert some errors into the program. This is called the error seeding method. Suppose that $n_1$ errors are inserted into the program and after a series of testing the programmer has found $n_2$ errors in which $n_{12}$ errors are inserted. Then if the inserted errors are as typical as the indigeneous (original) errors, all the formulas can be used. A special case is when all the inserted errors are found. In this case, $n_{12} = n_1$, which makes the point estimate of the unfound error to be 0. This is reasonable because it seems that all the errors have been found based on the fact that all the inserted errors have been discovered. To find a confidence upper bound for $N$, we use (8.3.4),

$$p = \binom{n_2}{n_1} \Big/ \binom{N}{n_1}$$

$$= \{n_2!(N - n_1)!\}/\{(n_2 - n_1)!N!\} \quad .$$

*Example 8.3.4:* One hundred errors were inserted into a program and after a long test, all the inserted errors are found with 1 indigeneous error. Find a 95% upper bound for the remaining errors.

*Solution:* $n_{12} = n_1 = 100, \qquad n_2 = 101$

$$p = \{101!(N - 100)!\}/N! \ .$$

For $N = 102$, $p = 2/102 < 0.05$. Thus, 1 remaining error is the upper 95% confidence bound.

**Exercise 8**

8.1 (i) One number is randomly picked from $N$ unequal numbers. What is the probability that it is the largest?

(ii) Now $n$ numbers are randomly picked without replacement, what is the probability that they contain the largest number?

8.2 In the previous exercise, let the numbers picked be $x(1), x(2), \ldots, x(n)$. A new number $x(n+1)$ is now picked from the remaining numbers. What is the probability that it is smaller than the second largest number of $x(1), \ldots, x(n)$? Use this idea to construct an algorithm that will find the largest two numbers in a data set of size $N$ with less than $(N-1)+(N-2)$ comparisons, on the average, for large $N$.

8.3 What is the chance that we get into the worst case situation in quicksort?

8.4 Should the quicksort algorithm be modified if we know that there are a lot of ties?

8.5 Can we simply use the first datum in each group as the fence in quicksort? What is the danger?

8.6 (i) A software was tested 1678 times without failure. Find a 99% confidence upper bound for its failure rate.

(ii) Fifty seeding errors are inserted into a piece of software and the software is tested by random inputs. After many trials, all the fifty inserted errors are found without any other errors. What kind of confidence do you have if you claim that the software contains no errors?

8.7 Some software was given to two testers. One found 35 errors and the other found 65 and there were 27 errors in common. Give a reasonable estimate of the undetected errors.

8.8 (*Programming exercise*) An urn contains 100 black balls and 10 white balls. We will draw balls without replacement until we get 95 black balls. Let $X$ = number of white balls drawn during this process. Write a simulation program to estimate the distribution of $X$.

# REFERENCES AND FURTHER STUDIES


1. Afifi, A. A., and Azen, S. P. (1979), *Statistical Analysis — A Computer Oriented Approach,* Academic Press, New York.

2. Alvarez, A. R., Welter, D. J., and Johnson, M. (1983), "Problem Solving in the IC Industry Through Applied Statistics: Comparing Two Processes," *Solid State Technology,* pp. 127-133.

3. Box, G. E. P. and Muller, M. E. (1958), "A Note on Generating of Normal Deviates," *Ann. Math. Stat.,* pp. 610-611.

4. David, H. A. (1981), *Order Statistics,* John Wiley & Sons, New York.

5. Duran, J. W., and Wiorkowski, J. J. (1980), "Quantify Software Validity by Sampling," *IEEE Transactions on Reliability,* pp. 141-144.

6. Edgington, E. S. (1980), *Randomization Tests,* Marcel Dekker, New York.

7. Efron, B. (1982), *The Jackknife, the Bootstrap and Other Resampling Plans,* SIAM Press, Philadelphia.

8. Feller, W. (1957), *An Introduction to Probability Theory and Its Applications,* Vol. I, John Wiley & Sons, New York.

9. Goodman, L. A. (1952), "Serial Number Analysis," *Journal of American Statistical Association,* pp. 622-634.

10. Gay, F. A. (1978), "Evaluation of Maintenance Software in Real-Time System," *IEEE Transactions on Computers,* pp. 567-582.

11. Huber, P. J. (1977), *Robust Statistical Procedures,* SIAM Press, Philadelphia.

12. Johnson, N. L. and Kotz, S. (1969), *Discrete Distributions,* John Wiley & Sons, New York.

13. Johnson, N. L. and Kotz, S. (1970), *Continuous Univariate Distributions-1,* John Wiley & Sons, New York.

14. Kennedy, W. J., and Gentle, J. E. (1980), *Statistical Computing,* Marcel DEkker, New York.

15. McClave, J. T., and Dietrich, F. H. (1985), *Statistics,* Dellen, San Francisco.

16. Mendenhall, W., Scheaffer, R. L., and Wackerly, D. (1984), *Mathematical Statistics with Applications,* Duxbury Press, North Scituate, Mass.

17. Myers, G. J. (1976), *Software Reliability, Principles and Practices,* John Wiley & Sons, New York.

18. Rubinstein, R. Y. (1981), *Simulation and the Monte Carlo Method,* John Wiley & Sons, New York.

19. Scheaffer, R. L. and McClave. J. T. (1982), *Statistics for Engineers,* Duxbury Press, Boston.

20. Trivedi, K. S. (1982), *Probability and Statistics with Reliability, Queuing, and Computer Science Applications,* Prentice-Hall, New Jersey.

21. Vitter, J. S. (1983), "Optimal Algorithm for Random Sampling Problems," 1983 *IEEE Fundamental of Computers,* pp. 65-75.

Table A.1 The $z$-table.

$$F(x) = \int_{-\infty}^{x} \frac{1}{\sqrt{2\pi}} e^{-\frac{1}{2}t^2}\, dt$$

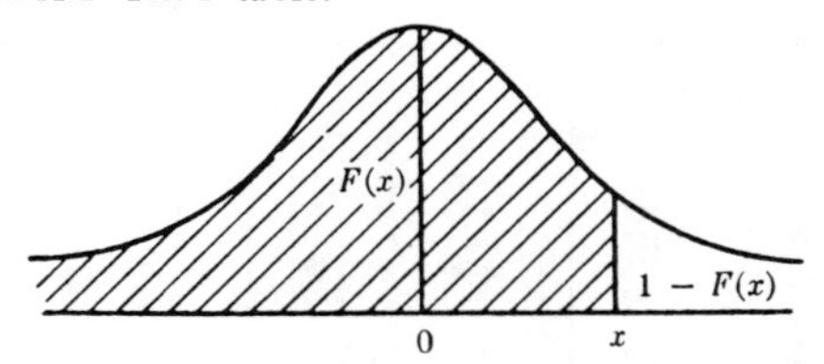


| $x$ | $F(x)$ | $1 - F(x)$ | $x$ | $F(x)$ | $1 - F(x)$ | $x$ | $F(x)$ | $1 - F(x)$ |
|---|---|---|---|---|---|---|---|---|
| .00 | .5000 | .5000 | .50 | .6915 | .3085 | 1.00 | .8413 | .1587 |
| .01 | .5040 | .4960 | .51 | .6950 | .3050 | 1.01 | .8438 | .1562 |
| .02 | .5080 | .4920 | .52 | .6985 | .3015 | 1.02 | .8461 | .1539 |
| .03 | .5120 | .4880 | .53 | .7019 | .2981 | 1.03 | .8485 | .1515 |
| .04 | .5160 | .4840 | .54 | .7054 | .2946 | 1.04 | .8508 | .1492 |
| .05 | .5199 | .4801 | .55 | .7088 | .2912 | 1.05 | .8531 | .1469 |
| .06 | .5239 | .4761 | .56 | .7123 | .2877 | 1.06 | .8554 | .1446 |
| .07 | .5279 | .4721 | .57 | .7157 | .2843 | 1.07 | .8577 | .1423 |
| .08 | .5319 | .4681 | .58 | .7190 | .2810 | 1.08 | .8599 | .1401 |
| .09 | .5359 | .4641 | .59 | .7224 | .2776 | 1.09 | .8621 | .1379 |
| .10 | .5398 | .4602 | .60 | .7257 | .2743 | 1.10 | .8643 | .1357 |
| .11 | .5438 | .4562 | .61 | .7291 | .2709 | 1.11 | .8665 | .1335 |
| .12 | .5478 | .4522 | .62 | .7324 | .2676 | 1.12 | .8686 | .1314 |
| .13 | .5517 | .4483 | .63 | .7357 | .2643 | 1.13 | .8708 | .1292 |
| .14 | .5557 | .4443 | .64 | .7389 | .2611 | 1.14 | .8729 | .1271 |
| .15 | .5596 | .4404 | .65 | .7422 | .2578 | 1.15 | .8749 | .1251 |
| .16 | .5636 | .4364 | .66 | .7454 | .2546 | 1.16 | .8770 | .1230 |
| .17 | .5675 | .4325 | .67 | .7486 | .2514 | 1.17 | .8790 | .1210 |
| .18 | .5714 | .4286 | .68 | .7517 | .2483 | 1.18 | .8810 | .1190 |
| .19 | .5753 | .4247 | .69 | .7549 | .2451 | 1.19 | .8830 | .1170 |
| .20 | .5793 | .4207 | .70 | .7580 | .2420 | 1.20 | .8849 | .1151 |
| .21 | .5832 | .4168 | .71 | .7611 | .2389 | 1.21 | .8869 | .1131 |
| .22 | .5871 | .4129 | .72 | .7642 | .2358 | 1.22 | .8888 | .1112 |
| .23 | .5910 | .4090 | .73 | .7673 | .2327 | 1.23 | .8907 | .1093 |
| .24 | .5948 | .4052 | .74 | .7704 | .2296 | 1.24 | .8925 | .1075 |
| .25 | .5987 | .4013 | .75 | .7734 | .2266 | 1.25 | .8944 | .1056 |
| .26 | .6026 | .3974 | .76 | .7764 | .2236 | 1.26 | .8962 | .1038 |
| .27 | .6064 | .3936 | .77 | .7794 | .2206 | 1.27 | .8980 | .1020 |
| .28 | .6103 | .3897 | .78 | .7823 | .2177 | 1.28 | .8997 | .1003 |
| .29 | .6141 | .3859 | .79 | .7852 | .2148 | 1.29 | .9015 | .0985 |
| .30 | .6179 | .3821 | .80 | .7881 | .2119 | 1.30 | .9032 | .0968 |
| .31 | .6217 | .3783 | .81 | .7910 | .2090 | 1.31 | .9049 | .0951 |
| .32 | .6255 | .3745 | .82 | .7939 | .2061 | 1.32 | .9066 | .0934 |
| .33 | .6293 | .3707 | .83 | .7967 | .2033 | 1.33 | .9082 | .0918 |
| .34 | .6331 | .3669 | .84 | .7995 | .2005 | 1.34 | .9099 | .0901 |
| .35 | .6368 | .3632 | .85 | .8023 | .1977 | 1.35 | .9115 | .0885 |
| .36 | .6406 | .3594 | .86 | .8051 | .1949 | 1.36 | .9131 | .0869 |
| .37 | .6443 | .3557 | .87 | .8078 | .1922 | 1.37 | .9147 | .0853 |
| .38 | .6480 | .3520 | .88 | .8106 | .1894 | 1.38 | .9162 | .0838 |
| .39 | .6517 | .3483 | .89 | .8133 | .1867 | 1.39 | .9177 | .0823 |
| .40 | .6554 | .3446 | .90 | .8159 | .1841 | 1.40 | .9192 | .0808 |
| .41 | .6591 | .3409 | .91 | .8186 | .1814 | 1.41 | .9207 | .0793 |
| .42 | .6628 | .3372 | .92 | .8212 | .1788 | 1.42 | .9222 | .0778 |
| .43 | .6664 | .3336 | .93 | .8238 | .1762 | 1.43 | .9236 | .0764 |
| .44 | .6700 | .3300 | .94 | .8264 | .1736 | 1.44 | .9251 | .0749 |
| .45 | .6736 | .3264 | .95 | .8289 | .1711 | 1.45 | .9265 | .0735 |
| .46 | .6772 | .3228 | .96 | .8315 | .1685 | 1.46 | .9279 | .0721 |
| .47 | .6808 | .3192 | .97 | .8340 | .1660 | 1.47 | .9292 | .0708 |
| .48 | .6844 | .3156 | .98 | .8365 | .1635 | 1.48 | .9306 | .0694 |
| .49 | .6879 | .3121 | .99 | .8389 | .1611 | 1.49 | .9319 | .0681 |
| .50 | .6915 | .3085 | 1.00 | .8413 | .1587 | 1.50 | .9332 | .0668 |

Table A.1 (continued)

| $x$ | $F(x)$ | $1 - F(x)$ | $x$ | $F(x)$ | $1 - F(x)$ | $x$ | $F(x)$ | $1 - F(x)$ |
|---|---|---|---|---|---|---|---|---|
| 1.50 | .9332 | .0668 | 2.00 | .9773 | .0227 | 2.50 | .9938 | .0062 |
| 1.51 | .9345 | .0655 | 2.01 | .9778 | .0222 | 2.51 | .9940 | .0060 |
| 1.52 | .9357 | .0643 | 2.02 | .9783 | .0217 | 2.52 | .9941 | .0059 |
| 1.53 | .9370 | .0630 | 2.03 | .9788 | .0212 | 2.53 | .9943 | .0057 |
| 1.54 | .9382 | .0618 | 2.04 | .9793 | .0207 | 2.54 | .9945 | .0055 |
| 1.55 | .9394 | .0606 | 2.05 | .9798 | .0202 | 2.55 | .9946 | .0054 |
| 1.56 | .9406 | .0594 | 2.06 | .9803 | .0197 | 2.56 | .9948 | .0052 |
| 1.57 | .9418 | .0582 | 2.07 | .9808 | .0192 | 2.57 | .9949 | .0051 |
| 1.58 | .9429 | .0571 | 2.08 | .9812 | .0188 | 2.58 | .9951 | .0049 |
| 1.59 | .9441 | .0559 | 2.09 | .9817 | .0183 | 2.59 | .9952 | .0048 |
| 1.60 | .9452 | .0548 | 2.10 | .9821 | .0179 | 2.60 | .9953 | .0047 |
| 1.61 | .9463 | .0537 | 2.11 | .9826 | .0174 | 2.61 | .9955 | .0045 |
| 1.62 | .9474 | .0526 | 2.12 | .9830 | .0170 | 2.62 | .9956 | .0044 |
| 1.63 | .9484 | .0516 | 2.13 | .9834 | .0166 | 2.63 | .9957 | .0043 |
| 1.64 | .9495 | .0505 | 2.14 | .9838 | .0162 | 2.64 | .9959 | .0041 |
| 1.65 | .9505 | .0495 | 2.15 | .9842 | .0158 | 2.65 | .9960 | .0040 |
| 1.66 | .9515 | .0485 | 2.16 | .9846 | .0154 | 2.66 | .9961 | .0039 |
| 1.67 | .9525 | .0475 | 2.17 | .9850 | .0150 | 2.67 | .9962 | .0038 |
| 1.68 | .9535 | .0465 | 2.18 | .9854 | .0146 | 2.68 | .9963 | .0037 |
| 1.69 | .9545 | .0455 | 2.19 | .9857 | .0143 | 2.69 | .9964 | .0036 |
| 1.70 | .9554 | .0446 | 2.20 | .9861 | .0139 | 2.70 | .9965 | .0035 |
| 1.71 | .9564 | .0436 | 2.21 | .9864 | .0136 | 2.71 | .9966 | .0034 |
| 1.72 | .9573 | .0427 | 2.22 | .9868 | .0132 | 2.72 | .9967 | .0033 |
| 1.73 | .9582 | .0418 | 2.23 | .9871 | .0129 | 2.73 | .9968 | .0032 |
| 1.74 | .9591 | .0409 | 2.24 | .9875 | .0125 | 2.74 | .9969 | .0031 |
| 1.75 | .9599 | .0401 | 2.25 | .9878 | .0122 | 2.75 | .9970 | .0030 |
| 1.76 | .9608 | .0392 | 2.26 | .9881 | .0119 | 2.76 | .9971 | .0029 |
| 1.77 | .9616 | .0384 | 2.27 | .9884 | .0116 | 2.77 | .9972 | .0028 |
| 1.78 | .9625 | .0375 | 2.28 | .9887 | .0113 | 2.78 | .9973 | .0027 |
| 1.79 | .9633 | .0367 | 2.29 | .9890 | .0110 | 2.79 | .9974 | .0026 |
| 1.80 | .9641 | .0359 | 2.30 | .9893 | .0107 | 2.80 | .9974 | .0026 |
| 1.81 | .9649 | .0351 | 2.31 | .9896 | .0104 | 2.81 | .9975 | .0025 |
| 1.82 | .9656 | .0344 | 2.32 | .9898 | .0102 | 2.82 | .9976 | .0024 |
| 1.83 | .9664 | .0336 | 2.33 | .9901 | .0099 | 2.83 | .9977 | .0023 |
| 1.84 | .9671 | .0329 | 2.34 | .9904 | .0096 | 2.84 | .9977 | .0023 |
| 1.85 | .9678 | .0322 | 2.35 | .9906 | .0094 | 2.85 | .9978 | .0022 |
| 1.86 | .9686 | .0314 | 2.36 | .9909 | .0091 | 2.86 | .9979 | .0021 |
| 1.87 | .9693 | .0307 | 2.37 | .9911 | .0089 | 2.87 | .9979 | .0021 |
| 1.88 | .9699 | .0301 | 2.38 | .9913 | .0087 | 2.88 | .9980 | .0020 |
| 1.89 | .9706 | .0294 | 2.39 | .9916 | .0084 | 2.89 | .9981 | .0019 |
| 1.90 | .9713 | .0287 | 2.40 | .9918 | .0082 | 2.90 | .9981 | .0019 |
| 1.91 | .9719 | .0281 | 2.41 | .9920 | .0080 | 2.91 | .9982 | .0018 |
| 1.92 | .9726 | .0274 | 2.42 | .9922 | .0078 | 2.92 | .9982 | .0018 |
| 1.93 | .9732 | .0268 | 2.43 | .9925 | .0075 | 2.93 | .9983 | .0017 |
| 1.94 | .9738 | .0262 | 2.44 | .9927 | .0073 | 2.94 | .9984 | .0016 |
| 1.95 | .9744 | .0256 | 2.45 | .9929 | .0071 | 2.95 | .9984 | .0016 |
| 1.96 | .9750 | .0250 | 2.46 | .9931 | .0069 | 2.96 | .9985 | .0015 |
| 1.97 | .9756 | .0244 | 2.47 | .9932 | .0068 | 2.97 | .9985 | .0015 |
| 1.98 | .9761 | .0239 | 2.48 | .9934 | .0066 | 2.98 | .9986 | .0014 |
| 1.99 | .9767 | .0233 | 2.49 | .9936 | .0064 | 2.99 | .9986 | .0014 |
| 2.00 | .9772 | .0228 | 2.50 | .9938 | .0062 | 3.00 | .9987 | .0013 |

Table A.1 (continued)

| $z$ | $F(z)$ | $1 - F(z)$ | $z$ | $F(z)$ | $1 - F(z)$ |
|---|---|---|---|---|---|
| 3.00 | .9987 | .0013 | 3.50 | .9998 | .0002 |
| 3.01 | .9987 | .0013 | 3.51 | .9998 | .0002 |
| 3.02 | .9987 | .0013 | 3.52 | .9998 | .0002 |
| 3.03 | .9988 | .0012 | 3.53 | .9998 | .0002 |
| 3.04 | .9988 | .0012 | 3.54 | .9998 | .0002 |
| 3.05 | .9989 | .0011 | 3.55 | .9998 | .0002 |
| 3.06 | .9989 | .0011 | 3.56 | .9998 | .0002 |
| 3.07 | .9989 | .0011 | 3.57 | .9998 | .0002 |
| 3.08 | .9990 | .0010 | 3.58 | .9998 | .0002 |
| 3.09 | .9990 | .0010 | 3.59 | .9998 | .0002 |
| 3.10 | .9990 | .0010 | 3.60 | .9998 | .0002 |
| 3.11 | .9991 | .0009 | 3.61 | .9998 | .0002 |
| 3.12 | .9991 | .0009 | 3.62 | .9999 | .0001 |
| 3.13 | .9991 | .0009 | 3.63 | .9999 | .0001 |
| 3.14 | .9992 | .0008 | 3.64 | .9999 | .0001 |
| 3.15 | .9992 | .0008 | 3.65 | .9999 | .0001 |
| 3.16 | .9992 | .0008 | 3.66 | .9999 | .0001 |
| 3.17 | .9992 | .0008 | 3.67 | .9999 | .0001 |
| 3.18 | .9993 | .0007 | 3.68 | .9999 | .0001 |
| 3.19 | .9993 | .0007 | 3.69 | .9999 | .0001 |
| 3.20 | .9993 | .0007 | 3.70 | .9999 | .0001 |
| 3.21 | .9993 | .0007 | 3.71 | .9999 | .0001 |
| 3.22 | .9994 | .0006 | 3.72 | .9999 | .0001 |
| 3.23 | .9994 | .0006 | 3.73 | .9999 | .0001 |
| 3.24 | .9994 | .0006 | 3.74 | .9999 | .0001 |
| 3.25 | .9994 | .0006 | 3.75 | .9999 | .0001 |
| 3.26 | .9994 | .0006 | 3.76 | .9999 | .0001 |
| 3.27 | .9995 | .0005 | 3.77 | .9999 | .0001 |
| 3.28 | .9995 | .0005 | 3.78 | .9999 | .0001 |
| 3.29 | .9995 | .0005 | 3.79 | .9999 | .0001 |
| 3.30 | .9995 | .0005 | 3.80 | .9999 | .0001 |
| 3.31 | .9995 | .0005 | 3.81 | .9999 | .0001 |
| 3.32 | .9995 | .0005 | 3.82 | .9999 | .0001 |
| 3.33 | .9996 | .0004 | 3.83 | .9999 | .0001 |
| 3.34 | .9996 | .0004 | 3.84 | .9999 | .0001 |
| 3.35 | .9996 | .0004 | 3.85 | .9999 | .0001 |
| 3.36 | .9996 | .0004 | 3.86 | .9999 | .0001 |
| 3.37 | .9996 | .0004 | 3.87 | .9999 | .0001 |
| 3.38 | .9996 | .0004 | 3.88 | .9999 | .0001 |
| 3.39 | .9997 | .0003 | 3.89 | 1.0000 | .0000 |
| 3.40 | .9997 | .0003 | 3.90 | 1.0000 | .0000 |
| 3.41 | .9997 | .0003 | 3.91 | 1.0000 | .0000 |
| 3.42 | .9997 | .0003 | 3.92 | 1.0000 | .0000 |
| 3.43 | .9997 | .0003 | 3.93 | 1.0000 | .0000 |
| 3.44 | .9997 | .0003 | 3.94 | 1.0000 | .0000 |
| 3.45 | .9997 | .0003 | 3.95 | 1.0000 | .0000 |
| 3.46 | .9997 | .0003 | 3.96 | 1.0000 | .0000 |
| 3.47 | .9997 | .0003 | 3.97 | 1.0000 | .0000 |
| 3.48 | .9997 | .0003 | 3.98 | 1.0000 | .0000 |
| 3.49 | .9998 | .0002 | 3.99 | 1.0000 | .0000 |
| 3.50 | .9998 | .0002 | 4.00 | 1.0000 | .0000 |

Table A.2

***Student's t-Distribution***

**PERCENTAGE POINTS, STUDENTS *t*-DISTRIBUTION**

$$F(t) = \int_{-\infty}^{t} \frac{\Gamma\left(\frac{n+1}{2}\right)}{\sqrt{n\pi}\,\Gamma\left(\frac{n}{2}\right)} \left(1 + \frac{x^2}{n}\right)^{-\frac{n+1}{2}} dx$$

| n \ F | .60 | .75 | .90 | .95 | .975 | .99 | .995 | .9995 |
|---|---|---|---|---|---|---|---|---|
| 1 | .325 | 1.000 | 3.078 | 6.314 | 12.706 | 31.821 | 63.657 | 636.619 |
| 2 | .289 | .816 | 1.886 | 2.920 | 4.303 | 6.965 | 9.925 | 31.598 |
| 3 | .277 | .765 | 1.638 | 2.353 | 3.182 | 4.541 | 5.841 | 12.924 |
| 4 | .271 | .741 | 1.533 | 2.132 | 2.776 | 3.747 | 4.604 | 8.610 |
| 5 | .267 | .727 | 1.476 | 2.015 | 2.571 | 3.365 | 4.032 | 6.869 |
| 6 | .265 | .718 | 1.440 | 1.943 | 2.447 | 3.143 | 3.707 | 5.959 |
| 7 | .263 | .711 | 1.415 | 1.895 | 2.365 | 2.998 | 3.499 | 5.408 |
| 8 | .262 | .706 | 1.397 | 1.860 | 2.306 | 2.896 | 3.355 | 5.041 |
| 9 | .261 | .703 | 1.383 | 1.833 | 2.262 | 2.821 | 3.250 | 4.781 |
| 10 | .260 | .700 | 1.372 | 1.812 | 2.228 | 2.764 | 3.169 | 4.587 |
| 11 | .260 | .697 | 1.363 | 1.796 | 2.201 | 2.718 | 3.106 | 4.437 |
| 12 | .259 | .695 | 1.356 | 1.782 | 2.179 | 2.681 | 3.055 | 4.318 |
| 13 | .259 | .694 | 1.350 | 1.771 | 2.160 | 2.650 | 3.012 | 4.221 |
| 14 | .258 | .692 | 1.345 | 1.761 | 2.145 | 2.624 | 2.977 | 4.140 |
| 15 | .258 | .691 | 1.341 | 1.753 | 2.131 | 2.602 | 2.947 | 4.073 |
| 16 | .258 | .690 | 1.337 | 1.746 | 2.120 | 2.583 | 2.921 | 4.015 |
| 17 | .257 | .689 | 1.333 | 1.740 | 2.110 | 2.567 | 2.898 | 3.965 |
| 18 | .257 | .688 | 1.330 | 1.734 | 2.101 | 2.552 | 2.878 | 3.922 |
| 19 | .257 | .688 | 1.328 | 1.729 | 2.093 | 2.539 | 2.861 | 3.883 |
| 20 | .257 | .687 | 1.325 | 1.725 | 2.086 | 2.528 | 2.845 | 3.850 |
| 21 | .257 | .686 | 1.323 | 1.721 | 2.080 | 2.518 | 2.831 | 3.819 |
| 22 | .256 | .686 | 1.321 | 1.717 | 2.074 | 2.508 | 2.819 | 3.792 |
| 23 | .256 | .685 | 1.319 | 1.714 | 2.069 | 2.500 | 2.807 | 3.767 |
| 24 | .256 | .685 | 1.318 | 1.711 | 2.064 | 2.492 | 2.797 | 3.745 |
| 25 | .256 | .684 | 1.316 | 1.708 | 2.060 | 2.485 | 2.787 | 3.725 |
| 26 | .256 | .684 | 1.315 | 1.706 | 2.056 | 2.479 | 2.779 | 3.707 |
| 27 | .256 | .684 | 1.314 | 1.703 | 2.052 | 2.473 | 2.771 | 3.690 |
| 28 | .256 | .683 | 1.313 | 1.701 | 2.048 | 2.467 | 2.763 | 3.674 |
| 29 | .256 | .683 | 1.311 | 1.699 | 2.045 | 2.462 | 2.756 | 3.659 |
| 30 | .256 | .683 | 1.310 | 1.697 | 2.042 | 2.457 | 2.750 | 3.646 |
| 40 | .255 | .681 | 1.303 | 1.684 | 2.021 | 2.423 | 2.704 | 3.551 |
| 60 | .254 | .679 | 1.296 | 1.671 | 2.000 | 2.390 | 2.660 | 3.460 |
| 120 | .254 | .677 | 1.289 | 1.658 | 1.980 | 2.358 | 2.617 | 3.373 |
| ∞ | .253 | .674 | 1.282 | 1.645 | 1.960 | 2.326 | 2.576 | 3.291 |

Table A.3

**PERCENTAGE POINTS, CHI-SQUARE DISTRIBUTION**

$$F(\chi^2) = \int_0^{\chi^2} \frac{1}{2^{\frac{n}{2}}\Gamma\left(\frac{n}{2}\right)} x^{\frac{n-2}{2}} e^{-\frac{x}{2}} dx$$

| n \ F | .005 | .010 | .025 | .050 | .100 | .250 | .500 | .750 | .900 | .950 | .975 | .990 | .995 |
|---|---|---|---|---|---|---|---|---|---|---|---|---|---|
| 1 | .0000393 | .000157 | .000982 | .00393 | .0158 | .102 | .455 | 1.32 | 2.71 | 3.84 | 5.02 | 6.63 | 7.88 |
| 2 | .0100 | .0201 | .0506 | .103 | .211 | .575 | 1.39 | 2.77 | 4.61 | 5.99 | 7.38 | 9.21 | 10.6 |
| 3 | .0717 | .115 | .216 | .352 | .584 | 1.21 | 2.37 | 4.11 | 6.25 | 7.81 | 9.35 | 11.3 | 12.8 |
| 4 | .207 | .297 | .484 | .711 | 1.06 | 1.92 | 3.36 | 5.39 | 7.78 | 9.49 | 11.1 | 13.3 | 14.9 |
| 5 | .412 | .554 | .831 | 1.15 | 1.61 | 2.67 | 4.35 | 6.63 | 9.24 | 11.1 | 12.8 | 15.1 | 16.7 |
| 6 | .676 | .872 | 1.24 | 1.64 | 2.20 | 3.45 | 5.35 | 7.84 | 10.6 | 12.6 | 14.4 | 16.8 | 18.5 |
| 7 | .989 | 1.24 | 1.69 | 2.17 | 2.83 | 4.25 | 6.35 | 9.04 | 12.0 | 14.1 | 16.0 | 18.5 | 20.3 |
| 8 | 1.34 | 1.65 | 2.18 | 2.73 | 3.49 | 5.07 | 7.34 | 10.2 | 13.4 | 15.5 | 17.5 | 20.1 | 22.0 |
| 9 | 1.73 | 2.09 | 2.70 | 3.33 | 4.17 | 5.90 | 8.34 | 11.4 | 14.7 | 16.9 | 19.0 | 21.7 | 23.6 |
| 10 | 2.16 | 2.56 | 3.25 | 3.94 | 4.87 | 6.74 | 9.34 | 12.5 | 16.0 | 18.3 | 20.5 | 23.2 | 25.2 |
| 11 | 2.60 | 3.05 | 3.82 | 4.57 | 5.58 | 7.58 | 10.3 | 13.7 | 17.3 | 19.7 | 21.9 | 24.7 | 26.8 |
| 12 | 3.07 | 3.57 | 4.40 | 5.23 | 6.30 | 8.44 | 11.3 | 14.8 | 18.5 | 21.0 | 23.3 | 26.2 | 28.3 |
| 13 | 3.57 | 4.11 | 5.01 | 5.89 | 7.04 | 9.30 | 12.3 | 16.0 | 19.8 | 22.4 | 24.7 | 27.7 | 29.8 |
| 14 | 4.07 | 4.66 | 5.63 | 6.57 | 7.79 | 10.2 | 13.3 | 17.1 | 21.1 | 23.7 | 26.1 | 29.1 | 31.3 |
| 15 | 4.60 | 5.23 | 6.26 | 7.26 | 8.55 | 11.0 | .4.3 | 18.2 | 22.3 | 25.0 | 27.5 | 30.6 | 32.8 |
| 16 | 5.14 | 5.81 | 6.91 | 7.96 | 9.31 | 11.9 | 15.3 | 19.4 | 23.5 | 26.3 | 28.8 | 32.0 | 34.3 |
| 17 | 5.70 | 6.41 | 7.56 | 8.67 | 10.1 | 12.8 | 16.3 | 20.5 | 24.8 | 27.6 | 30.2 | 33.4 | 35.7 |
| 18 | 6.26 | 7.01 | 8.23 | 9.39 | 10.9 | 13.7 | 17.3 | 21.6 | 26.0 | 28.9 | 31.5 | 34.8 | 37.2 |
| 19 | 6.84 | 7.63 | 8.91 | 10.1 | 11.7 | 14.6 | 18.3 | 22.7 | 27.2 | 30.1 | 32.9 | 36.2 | 38.6 |
| 20 | 7.43 | 8.26 | 9.59 | 10.9 | 12.4 | 15.5 | 19.3 | 23.8 | 28.4 | 31.4 | 34.2 | 37.6 | 40.0 |
| 21 | 8.03 | 8.90 | 10.3 | 11.6 | 13.2 | 16.3 | 20.3 | 24.9 | 29.6 | 32.7 | 35.5 | 38.9 | 41.4 |
| 22 | 8.64 | 9.54 | 11.0 | 12.3 | 14.0 | 17.2 | 21.3 | 26.0 | 30.8 | 33.9 | 36.8 | 40.3 | 42.8 |
| 23 | 9.26 | 10.2 | 11.7 | 13.1 | 14.8 | 18.1 | 22.3 | 27.1 | 32.0 | 35.2 | 38.1 | 41.6 | 44.2 |
| 24 | 9.89 | 10.9 | 12.4 | 13.8 | 15.7 | 19.0 | 23.3 | 28.2 | 33.2 | 36.4 | 39.4 | 43.0 | 45.6 |
| 25 | 10.5 | 11.5 | 13.1 | 14.6 | 16.5 | 19.9 | 24.3 | 29.3 | 34.4 | 37.7 | 40.6 | 44.3 | 46.9 |
| 26 | 11.2 | 12.2 | 13.8 | 15.4 | 17.3 | 20.8 | 25.3 | 30.4 | 35.6 | 38.9 | 41.9 | 45.6 | 48.3 |
| 27 | 11.8 | 12.9 | 14.6 | 16.2 | 18.1 | 21.7 | 26.3 | 31.5 | 36.7 | 40.1 | 43.2 | 47.0 | 49.6 |
| 28 | 12.5 | 13.6 | 15.3 | 16.9 | 18.9 | 22.7 | 27.3 | 32.6 | 37.9 | 41.3 | 44.5 | 48.3 | 51.0 |
| 29 | 13.1 | 14.3 | 16.0 | 17.7 | 19.8 | 23.6 | 28.3 | 33.7 | 39.1 | 42.6 | 45.7 | 49.6 | 52.3 |
| 30 | 13.8 | 15.0 | 16.8 | 18.5 | 20.6 | 24.5 | 29.3 | 34.8 | 40.3 | 43.8 | 47.0 | 50.9 | 53.7 |

Table A. 4

**PERCENTAGE POINTS, *F*-DISTRIBUTION**

$$F(F) = \int_0^F \frac{\Gamma\left(\frac{m+n}{2}\right)}{\Gamma\left(\frac{m}{2}\right)\Gamma\left(\frac{n}{2}\right)} m^{\frac{m}{2}} n^{\frac{n}{2}} x^{\frac{m}{2}-1} (n+mx)^{-\frac{m+n}{2}}\, dx = .90$$

| n \ m | 1 | 2 | 3 | 4 | 5 | 6 | 7 | 8 | 9 | 10 | 12 | 15 | 20 | 24 | 30 | 40 | 60 | 120 | ∞ |
|---|---|---|---|---|---|---|---|---|---|---|---|---|---|---|---|---|---|---|---|
| 1 | 39.86 | 49.50 | 53.59 | 55.83 | 57.24 | 58.20 | 58.91 | 59.44 | 59.86 | 60.19 | 60.71 | 61.22 | 61.74 | 62.00 | 62.26 | 62.53 | 62.79 | 63.06 | 63.33 |
| 2 | 8.53 | 9.00 | 9.16 | 9.24 | 9.29 | 9.33 | 9.35 | 9.37 | 9.38 | 9.39 | 9.41 | 9.42 | 9.44 | 9.45 | 9.46 | 9.47 | 9.47 | 9.48 | 9.49 |
| 3 | 5.54 | 5,46 | 5.39 | 5.34 | 5.31 | 5.28 | 5.27 | 5.25 | 5.24 | 5.23 | 5.22 | 5.20 | 5.18 | 5.18 | 5.17 | 5.16 | 5.15 | 5.14 | 5.13 |
| 4 | 4.54 | 4.32 | 4.19 | 4.11 | 4.05 | 4.01 | 3.98 | 3.95 | 3.94 | 3.92 | 3.90 | 3.87 | 3.84 | 3.83 | 3.82 | 3.80 | 3.79 | 3.78 | 3.76 |
| 5 | 4.06 | 3.78 | 3.62 | 3.52 | 3.45 | 3.40 | 3.37 | 3.34 | 3.32 | 3.30 | 3.27 | 3.24 | 3.21 | 3.19 | 3.17 | 3.16 | 3.14 | 3.12 | 3.10 |
| 6 | 3.78 | 3.46 | 3.29 | 3.18 | 3.11 | 3.05 | 3.01 | 2.98 | 2.96 | 2.94 | 2.90 | 2.87 | 2.84 | 2.82 | 2.80 | 2.78 | 2.76 | 2.74 | 2.72 |
| 7 | 3.59 | 3.26 | 3.07 | 2.96 | 2.88 | 2.83 | 2.78 | 2.75 | 2.72 | 2.70 | 2.67 | 2.63 | 2.59 | 2.58 | 2.56 | 2.54 | 2.51 | 2.49 | 2.47 |
| 8 | 3.46 | 3.11 | 2.92 | 2.81 | 2.73 | 2.67 | 2.62 | 2.59 | 2.56 | 2.54 | 2.50 | 2.46 | 2.42 | 2.40 | 2.38 | 2.36 | 2.34 | 2.32 | 2.29 |
| 9 | 3.36 | 3.01 | 2.81 | 2.69 | 2.61 | 2.55 | 2.51 | 2.47 | 2.44 | 2.42 | 2.38 | 2.34 | 2.30 | 2.28 | 2.25 | 2.23 | 2.21 | 2.18 | 2.16 |
| 10 | 3.29 | 2.92 | 2.73 | 2.61 | 2.52 | 2.46 | 2.41 | 2.38 | 2.35 | 2.32 | 2.28 | 2.24 | 2.20 | 2.18 | 2.16 | 2.13 | 2.11 | 2.08 | 2.06 |
| 11 | 3.23 | 2.86 | 2.66 | 2.54 | 2.45 | 2.39 | 2.34 | 2.30 | 2.27 | 2.25 | 2.21 | 2.17 | 2.12 | 2.10 | 2.08 | 2.05 | 2.03 | 2.00 | 1.97 |
| 12 | 3.18 | 2.81 | 2.61 | 2.48 | 2.39 | 2.33 | 2.28 | 2.24 | 2.21 | 2.19 | 2.15 | 2.10 | 2.06 | 2.04 | 2.01 | 1.99 | 1.96 | 1.93 | 1.90 |
| 13 | 3.14 | 2.76 | 2.56 | 2.43 | 2.35 | 2.28 | 2.23 | 2.20 | 2.16 | 2.14 | 2.10 | 2.05 | 2.01 | 1.98 | 1.96 | 1.93 | 1.90 | 1.88 | 1.85 |
| 14 | 3.10 | 2.73 | 2.52 | 2.39 | 2.31 | 2.24 | 2.19 | 2.15 | 2.12 | 2.10 | 2.05 | 2.01 | 1.96 | 1.94 | 1.91 | 1.89 | 1.86 | 1.83 | 1.80 |
| 15 | 3.07 | 2.70 | 2.49 | 2.36 | 2.27 | 2.21 | 2.16 | 2.12 | 2.09 | 2.06 | 2.02 | 1.97 | 1.92 | 1.90 | 1.87 | 1.85 | 1.82 | 1.79 | 1.76 |
| 16 | 3.05 | 2.67 | 2.46 | 2.33 | 2.24 | 2.18 | 2.13 | 2.09 | 2.06 | 2.03 | 1.99 | 1.94 | 1.89 | 1.87 | 1.84 | 1.81 | 1.78 | 1.75 | 1.72 |
| 17 | 3.03 | 2.64 | 2.44 | 2.31 | 2.22 | 2.15 | 2.10 | 2.06 | 2.03 | 2.00 | 1.96 | 1.91 | 1.86 | 1.84 | 1.81 | 1.78 | 1.75 | 1.72 | 1.69 |
| 18 | 3.01 | 2.62 | 2.42 | 2.29 | 2.20 | 2.13 | 2.08 | 2.04 | 2.00 | 1.98 | 1.93 | 1.89 | 1.84 | 1.81 | 1.78 | 1.75 | 1.72 | 1.69 | 1.66 |
| 19 | 2.99 | 2.61 | 2.40 | 2.27 | 2.18 | 2.11 | 2.06 | 2.02 | 1.98 | 1.96 | 1.91 | 1.86 | 1.81 | 1.79 | 1.76 | 1.73 | 1.70 | 1.67 | 1.63 |
| 20 | 2.97 | 2.59 | 2.38 | 2.25 | 2.16 | 2.09 | 2.04 | 2.00 | 1.96 | 1.94 | 1.89 | 1.84 | 1.79 | 1.77 | 1.74 | 1.71 | 1.68 | 1.64 | 1.61 |
| 21 | 2.96 | 2.57 | 2.36 | 2.23 | 2.14 | 2.08 | 2.02 | 1.98 | 1.95 | 1.92 | 1.87 | 1.83 | 1.78 | 1.75 | 1.72 | 1.69 | 1.66 | 1.62 | 1.59 |
| 22 | 2.95 | 2.56 | 2.35 | 2.22 | 2.13 | 2.06 | 2.01 | 1.97 | 1.93 | 1.90 | 1.86 | 1.81 | 1.76 | 1.73 | 1.70 | 1.67 | 1.64 | 1.60 | 1.57 |
| 23 | 2.94 | 2.55 | 2.34 | 2.21 | 2.11 | 2.05 | 1.99 | 1.95 | 1.92 | 1.89 | 1.84 | 1.80 | 1.74 | 1.72 | 1.69 | 1.66 | 1.62 | 1.59 | 1.55 |
| 24 | 2.93 | 2.54 | 2.33 | 2.19 | 2.10 | 2.04 | 1.98 | 1.94 | 1.91 | 1.88 | 1.83 | 1.78 | 1.73 | 1.70 | 1.67 | 1.64 | 1.61 | 1.57 | 1.53 |
| 25 | 2.92 | 2.53 | 2.32 | 2.18 | 2.09 | 2.02 | 1.97 | 1.93 | 1.89 | 1.87 | 1.82 | 1.77 | 1.72 | 1.69 | 1.66 | 1.63 | 1.59 | 1.56 | 1.62 |
| 26 | 2.91 | 2.52 | 2.31 | 2.17 | 2.08 | 2.01 | 1.96 | 1.92 | 1.88 | 1.86 | 1.81 | 1.76 | 1.71 | 1.68 | 1.65 | 1.61 | 1.58 | 1.54 | 1.50 |
| 27 | 2.90 | 2.51 | 2.30 | 2.17 | 2.07 | 2.00 | 1.95 | 1.91 | 1.87 | 1.85 | 1.80 | 1.75 | 1.70 | 1.67 | 1.64 | 1.60 | 1.57 | 1.53 | 1.49 |
| 28 | 2.89 | 2.50 | 2.29 | 2.16 | 2.06 | 2.00 | 1.94 | 1.90 | 1.87 | 1.84 | 1.79 | 1.74 | 1.69 | 1.66 | 1.63 | 1.59 | 1.56 | 1.52 | 1.48 |
| 29 | 2.89 | 2.50 | 2.28 | 2.15 | 2.06 | 1.99 | 1.93 | 1.89 | 1.86 | 1.83 | 1.78 | 1.73 | 1.68 | 1.65 | 1.62 | 1.58 | 1.55 | 1.51 | 1.47 |
| 30 | 2.88 | 2.49 | 2.28 | 2.14 | 2.05 | 1.98 | 1.93 | 1.88 | 1.85 | 1.82 | 1.77 | 1.72 | 1.67 | 1.64 | 1.61 | 1.57 | 1.54 | 1.50 | 1.46 |
| 40 | 2.84 | 2.44 | 2.23 | 2.09 | 2.00 | 1.93 | 1.87 | 1.83 | 1.79 | 1.76 | 1.71 | 1.66 | 1.61 | 1.57 | 1.54 | 1.51 | 1.47 | 1.42 | 1.38 |
| 60 | 2.79 | 2.39 | 2.18 | 2.04 | 1.95 | 1.87 | 1.82 | 1.77 | 1.74 | 1.71 | 1.66 | 1.60 | 1.54 | 1.51 | 1.48 | 1.44 | 1.40 | 1.35 | 1.29 |
| 120 | 2.75 | 2.35 | 2.13 | 1.99 | 1.90 | 1.82 | 1.77 | 1.72 | 1.68 | 1.65 | 1.60 | 1.55 | 1.48 | 1.45 | 1.41 | 1.37 | 1.32 | 1.26 | 1.19 |
| ∞ | 2.71 | 2.30 | 2.08 | 1.94 | 1.85 | 1.77 | 1.72 | 1.67 | 1.63 | 1.60 | 1.55 | 1.49 | 1.42 | 1.38 | 1.34 | 1.30 | 1.24 | 1.17 | 1.00 |

$F = \frac{s_1^2}{s_2^2} = \frac{S_1}{m} \Big/ \frac{S_2}{n}$, where $s_1^2 = S_1/m$ and $s_2^2 = S_2/n$ are independent mean squares estimating a common variance $\sigma^2$ and based on $m$ and $n$ degrees of freedom, respectively.

Table A.4 (continued)

**PERCENTAGE POINTS, $F$-DISTRIBUTION**

$$F(F) = \int_0^F \frac{\Gamma\left(\frac{m+n}{2}\right)}{\Gamma\left(\frac{m}{2}\right)\Gamma\left(\frac{n}{2}\right)} m^{\frac{m}{2}} n^{\frac{n}{2}} x^{\frac{m}{2}-1} (n+mx)^{-\frac{m+n}{2}} dx = .95$$

| $n$ \ $m$ | 1 | 2 | 3 | 4 | 5 | 6 | 7 | 8 | 9 | 10 | 12 | 15 | 20 | 24 | 30 | 40 | 60 | 120 | ∞ |
|---|---|---|---|---|---|---|---|---|---|---|---|---|---|---|---|---|---|---|---|
| 1 | 161.4 | 199.5 | 215.7 | 224.6 | 230.2 | 234.0 | 236.8 | 238.9 | 240.5 | 241.9 | 243.9 | 245.9 | 248.0 | 249.1 | 250.1 | 251.1 | 252.2 | 253.3 | 254.3 |
| 2 | 18.51 | 19.00 | 19.16 | 19.25 | 19.30 | 19.33 | 19.35 | 19.37 | 19.38 | 19.40 | 19.41 | 19.43 | 19.45 | 19.45 | 19.46 | 19.47 | 19.48 | 19.49 | 19.50 |
| 3 | 10.13 | 9.55 | 9.28 | 9.12 | 9.01 | 8.94 | 8.89 | 8.85 | 8.81 | 8.79 | 8.74 | 8.70 | 8.66 | 8.64 | 8.62 | 8.59 | 8.57 | 8.55 | 8.53 |
| 4 | 7.71 | 6.94 | 6.59 | 6.39 | 6.26 | 6.16 | 6.09 | 6.04 | 6.00 | 5.96 | 5.91 | 5.86 | 5.80 | 5.77 | 5.75 | 5.72 | 5.69 | 5.66 | 5.63 |
| 5 | 6.61 | 5.79 | 5.41 | 5.19 | 5.05 | 4.95 | 4.88 | 4.82 | 4.77 | 4.74 | 4.68 | 4.62 | 4.56 | 4.53 | 4.50 | 4.46 | 4.43 | 4.40 | 4.36 |
| 6 | 5.99 | 5.14 | 4.76 | 4.53 | 4.39 | 4.28 | 4.21 | 4.15 | 4.10 | 4.06 | 4.00 | 3.94 | 3.87 | 3.84 | 3.81 | 3.77 | 3.74 | 3.70 | 3.67 |
| 7 | 5.59 | 4.74 | 4.35 | 4.12 | 3.97 | 3.87 | 3.79 | 3.73 | 3.68 | 3.64 | 3.57 | 3.51 | 3.44 | 3.41 | 3.38 | 3.34 | 3.30 | 3.27 | 3.23 |
| 8 | 5.32 | 4.46 | 4.07 | 3.84 | 3.69 | 3.58 | 3.50 | 3.44 | 3.39 | 3.35 | 3.28 | 3.22 | 3.15 | 3.12 | 3.08 | 3.04 | 3.01 | 2.97 | 2.93 |
| 9 | 5.12 | 4.26 | 3.86 | 3.63 | 3.48 | 3.37 | 3.29 | 3.23 | 3.18 | 3.14 | 3.07 | 3.01 | 2.94 | 2.90 | 2.86 | 2.83 | 2.79 | 2.75 | 2.71 |
| 10 | 4.96 | 4.10 | 3.71 | 3.48 | 3.33 | 3.22 | 3.14 | 3.07 | 3.02 | 2.98 | 2.91 | 2.85 | 2.77 | 2.74 | 2.70 | 2.66 | 2.62 | 2.58 | 2.54 |
| 11 | 4.84 | 3.98 | 3.59 | 3.36 | 3.20 | 3.09 | 3.01 | 2.95 | 2.90 | 2.85 | 2.79 | 2.72 | 2.65 | 2.61 | 2.57 | 2.53 | 2.49 | 2.45 | 2.40 |
| 12 | 4.75 | 3.89 | 3.49 | 3.26 | 3.11 | 3.00 | 2.91 | 2.85 | 2.80 | 2.75 | 2.69 | 2.62 | 2.54 | 2.51 | 2.47 | 2.43 | 2.38 | 2.34 | 2.30 |
| 13 | 4.67 | 3.81 | 3.41 | 3.18 | 3.03 | 2.92 | 2.83 | 2.77 | 2.71 | 2.67 | 2.60 | 2.53 | 2.46 | 2.42 | 2.38 | 2.34 | 2.30 | 2.25 | 2.21 |
| 14 | 4.60 | 3.74 | 3.34 | 3.11 | 2.96 | 2.85 | 2.76 | 2.70 | 2.65 | 2.60 | 2.53 | 2.46 | 2.39 | 2.35 | 2.31 | 2.27 | 2.22 | 2.18 | 2.13 |
| 15 | 4.54 | 3.68 | 3.29 | 3.06 | 2.90 | 2.79 | 2.71 | 2.64 | 2.59 | 2.54 | 2.48 | 2.40 | 2.33 | 2.29 | 2.25 | 2.20 | 2.16 | 2.11 | 2.07 |
| 16 | 4.49 | 3.63 | 3.24 | 3.01 | 2.85 | 2.74 | 2.66 | 2.59 | 2.54 | 2.49 | 2.42 | 2.35 | 2.28 | 2.24 | 2.19 | 2.15 | 2.11 | 2.06 | 2.01 |
| 17 | 4.45 | 3.59 | 3.20 | 2.96 | 2.81 | 2.70 | 2.61 | 2.55 | 2.49 | 2.45 | 2.38 | 2.31 | 2.23 | 2.19 | 2.15 | 2.10 | 2.06 | 2.01 | 1.96 |
| 18 | 4.41 | 3.55 | 3.16 | 2.93 | 2.77 | 2.66 | 2.58 | 2.51 | 2.46 | 2.41 | 2.34 | 2.27 | 2.19 | 2.15 | 2.11 | 2.06 | 2.02 | 1.97 | 1.92 |
| 19 | 4.38 | 3.52 | 3.13 | 2.90 | 2.74 | 2.63 | 2.54 | 2.48 | 2.42 | 2.38 | 2.31 | 2.23 | 2.16 | 2.11 | 2.07 | 2.03 | 1.98 | 1.93 | 1.88 |
| 20 | 4.35 | 3.49 | 3.10 | 2.87 | 2.71 | 2.60 | 2.51 | 2.45 | 2.39 | 2.35 | 2.28 | 2.20 | 2.12 | 2.08 | 2.04 | 1.99 | 1.95 | 1.90 | 1.84 |
| 21 | 4.32 | 3.47 | 3.07 | 2.84 | 2.68 | 2.57 | 2.49 | 2.42 | 2.37 | 2.32 | 2.25 | 2.18 | 2.10 | 2.05 | 2.01 | 1.96 | 1.92 | 1.87 | 1.81 |
| 22 | 4.30 | 3.44 | 3.05 | 2.82 | 2.66 | 2.55 | 2.46 | 2.40 | 2.34 | 2.30 | 2.23 | 2.15 | 2.07 | 2.03 | 1.98 | 1.94 | 1.89 | 1.84 | 1.78 |
| 23 | 4.28 | 3.42 | 3.03 | 2.80 | 2.64 | 2.53 | 2.44 | 2.37 | 2.32 | 2.27 | 2.20 | 2.13 | 2.05 | 2.01 | 1.96 | 1.91 | 1.86 | 1.81 | 1.76 |
| 24 | 4.26 | 3.40 | 3.01 | 2.78 | 2.62 | 2.51 | 2.42 | 2.36 | 2.30 | 2.25 | 2.18 | 2.11 | 2.03 | 1.98 | 1.94 | 1.89 | 1.84 | 1.79 | 1.73 |
| 25 | 4.24 | 3.39 | 2.99 | 2.76 | 2.60 | 2.49 | 2.40 | 2.34 | 2.28 | 2.24 | 2.16 | 2.09 | 2.01 | 1.96 | 1.92 | 1.87 | 1.82 | 1.77 | 1.71 |
| 26 | 4.23 | 3.37 | 2.98 | 2.74 | 2.59 | 2.47 | 2.39 | 2.32 | 2.27 | 2.22 | 2.15 | 2.07 | 1.99 | 1.95 | 1.90 | 1.85 | 1.80 | 1.75 | 1.69 |
| 27 | 4.21 | 3.35 | 2.96 | 2.73 | 2.57 | 2.46 | 2.37 | 2.31 | 2.25 | 2.20 | 2.13 | 2.06 | 1.97 | 1.93 | 1.88 | 1.84 | 1.79 | 1.73 | 1.67 |
| 28 | 4.20 | 3.34 | 2.95 | 2.71 | 2.56 | 2.45 | 2.36 | 2.29 | 2.24 | 2.19 | 2.12 | 2.04 | 1.96 | 1.91 | 1.87 | 1.82 | 1.77 | 1.71 | 1.65 |
| 29 | 4.18 | 3.33 | 2.93 | 2.70 | 2.55 | 2.43 | 2.35 | 2.28 | 2.22 | 2.18 | 2.10 | 2.03 | 1.94 | 1.90 | 1.85 | 1.81 | 1.75 | 1.70 | 1.64 |
| 30 | 4.17 | 3.32 | 2.92 | 2.69 | 2.53 | 2.42 | 2.33 | 2.27 | 2.21 | 2.16 | 2.09 | 2.01 | 1.93 | 1.89 | 1.84 | 1.79 | 1.74 | 1.68 | 1.62 |
| 40 | 4.08 | 3.23 | 2.84 | 2.61 | 2.45 | 2.34 | 2.25 | 2.18 | 2.12 | 2.08 | 2.00 | 1.92 | 1.84 | 1.79 | 1.74 | 1.69 | 1.64 | 1.58 | 1.51 |
| 60 | 4.00 | 3.15 | 2.76 | 2.53 | 2.37 | 2.25 | 2.17 | 2.10 | 2.04 | 1.99 | 1.92 | 1.84 | 1.75 | 1.70 | 1.65 | 1.59 | 1.53 | 1.47 | 1.39 |
| 120 | 3.92 | 3.07 | 2.68 | 2.45 | 2.29 | 2.17 | 2.09 | 2.02 | 1.96 | 1.91 | 1.83 | 1.75 | 1.66 | 1.61 | 1.55 | 1.50 | 1.43 | 1.35 | 1.25 |
| ∞ | 3.84 | 3.00 | 2.60 | 2.37 | 2.21 | 2.10 | 2.01 | 1.94 | 1.88 | 1.83 | 1.75 | 1.67 | 1.57 | 1.52 | 1.46 | 1.39 | 1.32 | 1.22 | 1.00 |

$F = \frac{s_1^2}{s_2^2} = \frac{S_1}{m} \Big/ \frac{S_2}{n}$, where $s_1^2 = S_1/m$ and $s_2^2 = S_2/n$ are independent mean squares estimating a common variance $\sigma^2$ and based on $m$ and $n$ degrees of freedom, respectively.

Table A.4 (continued)

**PERCENTAGE POINTS, *F*-DISTRIBUTION**

$$F(F) = \int_0^F \frac{\Gamma\left(\frac{m+n}{2}\right)}{\Gamma\left(\frac{m}{2}\right)\Gamma\left(\frac{n}{2}\right)} m^{\frac{m}{2}} n^{\frac{n}{2}} x^{\frac{m}{2}-1} (n+mx)^{-\frac{m+n}{2}}\, dx = .99$$

| n \ m | 1 | 2 | 3 | 4 | 5 | 6 | 7 | 8 | 9 | 10 | 12 | 15 | 20 | 24 | 30 | 40 | 60 | 120 | ∞ |
|---|---|---|---|---|---|---|---|---|---|---|---|---|---|---|---|---|---|---|---|
| 1 | 4052 | 4999.5 | 5403 | 5625 | 5764 | 5859 | 5928 | 5982 | 6022 | 6056 | 6106 | 6157 | 6209 | 6235 | 6261 | 6287 | 6313 | 6339 | 6366 |
| 2 | 98.50 | 99.00 | 99.17 | 99.25 | 99.30 | 99.33 | 99.36 | 99.37 | 99.39 | 99.40 | 99.42 | 99.43 | 99.45 | 99.46 | 99.47 | 99.47 | 99.48 | 99.49 | 99.50 |
| 3 | 34.12 | 30.82 | 29.46 | 28.71 | 28.24 | 27.91 | 27.67 | 27.49 | 27.35 | 27.23 | 27.05 | 26.87 | 26.69 | 26.60 | 26.50 | 26.41 | 26.32 | 26.22 | 26.13 |
| 4 | 21.20 | 18.00 | 16.69 | 15.98 | 15.52 | 15.21 | 14.98 | 14.80 | 14.66 | 14.55 | 14.37 | 14.20 | 14.02 | 13.93 | 13.84 | 13.75 | 13.65 | 13.56 | 13.46 |
| 5 | 16.26 | 13.27 | 12.06 | 11.39 | 10.97 | 10.67 | 10.46 | 10.29 | 10.16 | 10.05 | 9.89 | 9.72 | 9.55 | 9.47 | 9.38 | 9.29 | 9.20 | 9.11 | 9.02 |
| 6 | 13.75 | 10.92 | 9.78 | 9.15 | 8.75 | 8.47 | 8.26 | 8.10 | 7.98 | 7.87 | 7.72 | 7.56 | 7.40 | 7.31 | 7.23 | 7.14 | 7.06 | 6.97 | 6.88 |
| 7 | 12.25 | 9.55 | 8.45 | 7.85 | 7.46 | 7.19 | 6.99 | 6.84 | 6.72 | 6.62 | 6.47 | 6.31 | 6.16 | 6.07 | 5.99 | 5.91 | 5.82 | 5.74 | 5.65 |
| 8 | 11.26 | 8.65 | 7.59 | 7.01 | 6.63 | 6.37 | 6.18 | 6.03 | 5.91 | 5.81 | 5.67 | 5.52 | 5.36 | 5.28 | 5.20 | 5.12 | 5.03 | 4.95 | 4.86 |
| 9 | 10.56 | 8.02 | 6.99 | 6.42 | 6.06 | 5.80 | 5.61 | 5.47 | 5.35 | 5.26 | 5.11 | 4.96 | 4.81 | 4.73 | 4.65 | 4.57 | 4.48 | 4.40 | 4.31 |
| 10 | 10.04 | 7.56 | 6.55 | 5.99 | 5.64 | 5.39 | 5.20 | 5.06 | 4.94 | 4.85 | 4.71 | 4.56 | 4.41 | 4.33 | 4.25 | 4.17 | 4.08 | 4.00 | 3.91 |
| 11 | 9.65 | 7.21 | 6.22 | 5.67 | 5.32 | 5.07 | 4.89 | 4.74 | 4.63 | 4.54 | 4.40 | 4.25 | 4.10 | 4.02 | 3.94 | 3.86 | 3.78 | 3.69 | 3.60 |
| 12 | 9.33 | 6.93 | 5.95 | 5.41 | 5.06 | 4.82 | 4.64 | 4.50 | 4.39 | 4.30 | 4.16 | 4.01 | 3.86 | 3.78 | 3.70 | 3.62 | 3.54 | 3.45 | 3.36 |
| 13 | 9.07 | 6.70 | 5.74 | 5.21 | 4.86 | 4.62 | 4.44 | 4.30 | 4.19 | 4.10 | 3.96 | 3.82 | 3.66 | 3.59 | 3.51 | 3.43 | 3.34 | 3.25 | 3.17 |
| 14 | 8.86 | 6.51 | 5.56 | 5.04 | 4.69 | 4.46 | 4.28 | 4.14 | 4.03 | 3.94 | 3.80 | 3.66 | 3.51 | 3.43 | 3.35 | 3.27 | 3.18 | 3.09 | 3.00 |
| 15 | 8.68 | 6.36 | 5.42 | 4.89 | 4.56 | 4.32 | 4.14 | 4.00 | 3.89 | 3.80 | 3.67 | 3.52 | 3.37 | 3.29 | 3.21 | 3.13 | 3.05 | 2.96 | 2.87 |
| 16 | 8.53 | 6.23 | 5.29 | 4.77 | 4.44 | 4.20 | 4.03 | 3.89 | 3.78 | 3.69 | 3.55 | 3.41 | 3.26 | 3.18 | 3.10 | 3.02 | 2.93 | 2.84 | 2.75 |
| 17 | 8.40 | 6.11 | 5.18 | 4.67 | 4.34 | 4.10 | 3.93 | 3.79 | 3.68 | 3.59 | 3.46 | 3.31 | 3.16 | 3.08 | 3.00 | 2.92 | 2.83 | 2.75 | 2.65 |
| 18 | 8.29 | 6.01 | 5.09 | 4.58 | 4.25 | 4.01 | 3.84 | 3.71 | 3.60 | 3.51 | 3.37 | 3.23 | 3.08 | 3.00 | 2.92 | 2.84 | 2.75 | 2.66 | 2.57 |
| 19 | 8.18 | 5.93 | 5.01 | 4.50 | 4.17 | 3.94 | 3.77 | 3.63 | 3.52 | 3.43 | 3.30 | 3.15 | 3.00 | 2.92 | 2.84 | 2.76 | 2.67 | 2.58 | 2.49 |
| 20 | 8.10 | 5.85 | 4.94 | 4.43 | 4.10 | 3.87 | 3.70 | 3.56 | 3.46 | 3.37 | 3.23 | 3.09 | 2.94 | 2.86 | 2.78 | 2.69 | 2.61 | 2.52 | 2.42 |
| 21 | 8.02 | 5.78 | 4.87 | 4.37 | 4.04 | 3.81 | 3.64 | 3.51 | 3.40 | 3.31 | 3.17 | 3.03 | 2.88 | 2.80 | 2.72 | 2.64 | 2.55 | 2.46 | 2.36 |
| 22 | 7.95 | 5.72 | 4.82 | 4.31 | 3.99 | 3.76 | 3.59 | 3.45 | 3.35 | 3.26 | 3.12 | 2.98 | 2.83 | 2.75 | 2.67 | 2.58 | 2.50 | 2.40 | 2.31 |
| 23 | 7.88 | 5.66 | 4.76 | 4.26 | 3.94 | 3.71 | 3.54 | 3.41 | 3.30 | 3.21 | 3.07 | 2.93 | 2.78 | 2.70 | 2.62 | 2.54 | 2.45 | 2.35 | 2.26 |
| 24 | 7.82 | 5.61 | 4.72 | 4.22 | 3.90 | 3.67 | 3.50 | 3.36 | 3.26 | 3.17 | 3.03 | 2.89 | 2.74 | 2.66 | 2.58 | 2.49 | 2.40 | 2.31 | 2.21 |
| 25 | 7.77 | 5.57 | 4.68 | 4.18 | 3.85 | 3.63 | 3.46 | 3.32 | 3.22 | 3.13 | 2.99 | 2.85 | 2.70 | 2.62 | 2.54 | 2.45 | 2.36 | 2.27 | 2.17 |
| 26 | 7.72 | 5.53 | 4.64 | 4.14 | 3.82 | 3.59 | 3.42 | 3.29 | 3.18 | 3.09 | 2.96 | 2.81 | 2.66 | 2.58 | 2.50 | 2.42 | 2.33 | 2.23 | 2.13 |
| 27 | 7.68 | 5.49 | 4.60 | 4.11 | 3.78 | 3.56 | 3.39 | 3.26 | 3.15 | 3.06 | 2.93 | 2.78 | 2.63 | 2.55 | 2.47 | 2.38 | 2.29 | 2.20 | 2.10 |
| 28 | 7.64 | 5.45 | 4.57 | 4.07 | 3.75 | 3.53 | 3.36 | 3.23 | 3.12 | 3.03 | 2.90 | 2.75 | 2.60 | 2.52 | 2.44 | 2.35 | 2.26 | 2.17 | 2.06 |
| 29 | 7.60 | 5.42 | 4.54 | 4.04 | 3.73 | 3.50 | 3.33 | 3.20 | 3.09 | 3.00 | 2.87 | 2.73 | 2.57 | 2.49 | 2.41 | 2.33 | 2.23 | 2.14 | 2.03 |
| 30 | 7.56 | 5.39 | 4.51 | 4.02 | 3.70 | 3.47 | 3.30 | 3.17 | 3.07 | 2.98 | 2.84 | 2.70 | 2.55 | 2.47 | 2.39 | 2.30 | 2.21 | 2.11 | 2.01 |
| 40 | 7.31 | 5.18 | 4.31 | 3.83 | 3.51 | 3.29 | 3.12 | 2.99 | 2.89 | 2.80 | 2.66 | 2.52 | 2.37 | 2.29 | 2.20 | 2.11 | 2.02 | 1.92 | 1.80 |
| 60 | 7.08 | 4.98 | 4.13 | 3.65 | 3.34 | 3.12 | 2.95 | 2.82 | 2.72 | 2.63 | 2.50 | 2.35 | 2.20 | 2.12 | 2.03 | 1.94 | 1.84 | 1.73 | 1.60 |
| 120 | 6.85 | 4.79 | 3.95 | 3.48 | 3.17 | 2.96 | 2.79 | 2.66 | 2.56 | 2.47 | 2.34 | 2.19 | 2.03 | 1.95 | 1.86 | 1.76 | 1.66 | 1.53 | 1.38 |
| ∞ | 6.63 | 4.61 | 3.78 | 3.32 | 3.02 | 2.80 | 2.64 | 2.51 | 2.41 | 2.32 | 2.18 | 2.04 | 1.88 | 1.79 | 1.70 | 1.59 | 1.47 | 1.32 | 1.00 |

$F = \frac{s_1^2}{s_2^2} = \frac{S_1}{m} \Big/ \frac{S_2}{n}$, where $s_1^2 = S_1/m$ and $s_2^2 = S_2/n$ are independent mean squares estimating a common variance $\sigma^2$ and based on $m$ and $n$ degrees of freedom, respectively.

Table A.5 The Binomial Table $p(x) = \binom{n}{x} \theta^x (1-\theta)^{n-x}$.

**INDIVIDUAL TERMS, BINOMIAL DISTRIBUTION**

| n | x | θ .05 | .10 | .15 | .20 | .25 | .30 | .35 | .40 | .45 | .50 |
|---|---|---|---|---|---|---|---|---|---|---|---|
| 1 | 0 | .9500 | .9000 | .8500 | .8000 | .7500 | .7000 | .6500 | .6000 | .5500 | .5000 |
| | 1 | .0500 | .1000 | .1500 | .2000 | .2500 | .3000 | .3500 | .4000 | .4500 | .5000 |
| 2 | 0 | .9025 | .8100 | .7225 | .6400 | .5625 | .4900 | .4225 | .3600 | .3025 | .2500 |
| | 1 | .0950 | .1800 | .2550 | .3200 | .3750 | .4200 | .4550 | .4800 | .4950 | .5000 |
| | 2 | .0025 | .0100 | .0225 | .0400 | .0625 | .0900 | .1225 | .1600 | .2025 | .2500 |
| 3 | 0 | .8574 | .7290 | .6141 | .5120 | .4219 | .3430 | .2746 | .2160 | .1664 | .1250 |
| | 1 | .1354 | .2430 | .3251 | .3840 | .4219 | .4410 | .4436 | .4320 | .4084 | .3750 |
| | 2 | .0071 | .0270 | .0574 | .0960 | .1406 | .1890 | .2389 | .2880 | .3341 | .3750 |
| | 3 | .0001 | .0010 | .0034 | .0080 | .0156 | .0270 | .0429 | .0640 | .0911 | .1250 |
| 4 | 0 | .8145 | .6561 | .5220 | .4096 | .3164 | .2401 | .1785 | .1296 | .0915 | .0625 |
| | 1 | .1715 | .2916 | .3685 | .4096 | .4219 | .4116 | .3845 | .3456 | .2995 | .2500 |
| | 2 | .0135 | .0486 | .0975 | .1536 | .2109 | .2646 | .3105 | .3456 | .3675 | .3750 |
| | 3 | .0005 | .0036 | .0115 | .0256 | .0469 | .0756 | .1115 | .1536 | .2005 | .2500 |
| | 4 | .0000 | .0001 | .0005 | .0016 | .0039 | .0081 | .0150 | .0256 | .0410 | .0625 |
| 5 | 0 | .7738 | .5905 | .4437 | .3277 | .2373 | .1681 | .1160 | .0778 | .0503 | .0312 |
| | 1 | .2036 | .3280 | .3915 | .4096 | .3955 | .3602 | .3124 | .2592 | .2059 | .1562 |
| | 2 | .0214 | .0729 | .1382 | .2048 | .2637 | .3087 | .3364 | .3456 | .3369 | .3125 |
| | 3 | .0011 | .0081 | .0244 | .0512 | .0879 | .1323 | .1811 | .2304 | .2757 | .3125 |
| | 4 | .0000 | .0004 | .0022 | .0064 | .0146 | .0284 | .0488 | .0768 | .1128 | .1562 |
| | 5 | .0000 | .0000 | .0001 | .0003 | .0010 | .0024 | .0053 | .0102 | .0185 | .0312 |
| 6 | 0 | .7351 | .5314 | .3771 | .2621 | .1780 | .1176 | .0754 | .0467 | .0277 | .0156 |
| | 1 | .2321 | .3543 | .3993 | .3932 | .3560 | .3025 | .2437 | .1866 | .1359 | .0938 |
| | 2 | .0305 | .0984 | .1762 | .2458 | .2966 | .3241 | .3280 | .3110 | .2780 | .2344 |
| | 3 | .0021 | .0146 | .0415 | .0819 | .1318 | .1852 | .2355 | .2765 | .3032 | .3125 |
| | 4 | .0001 | .0012 | .0055 | .0154 | .0330 | .0595 | .0951 | .1382 | .1861 | .2344 |
| | 5 | .0000 | .0001 | .0004 | .0015 | .0044 | .0102 | .0205 | .0369 | .0609 | .0938 |
| | 6 | .0000 | .0000 | .0000 | .0001 | .0002 | .0007 | .0018 | .0041 | .0083 | .0156 |
| 7 | 0 | .6983 | .4783 | .3206 | .2097 | .1335 | .0824 | .0490 | .0280 | .0152 | .0078 |
| | 1 | .2573 | .3720 | .3960 | .3670 | .3115 | .2471 | .1848 | .1306 | .0872 | .0547 |
| | 2 | .0406 | .1240 | .2097 | .2753 | .3115 | .3177 | .2985 | .2613 | .2140 | .1641 |
| | 3 | .0036 | .0230 | .0617 | .1147 | .1730 | .2269 | .2679 | .2903 | .2918 | .2734 |
| | 4 | .0002 | .0026 | .0109 | .0287 | .0577 | .0972 | .1442 | .1935 | .2388 | .2734 |
| | 5 | .0000 | .0002 | .0012 | .0043 | .0115 | .0250 | .0466 | .0774 | .1172 | .1641 |
| | 6 | .0000 | .0000 | .0001 | .0004 | .0013 | .0036 | .0084 | .0172 | .0320 | .0547 |
| | 7 | .0000 | .0000 | .0000 | .0000 | .0001 | .0002 | .0006 | .0016 | .0037 | .0078 |
| 8 | 0 | .6634 | .4305 | .2725 | .1678 | .1001 | .0576 | .0319 | .0168 | .0084 | .0039 |
| | 1 | .2793 | .3826 | .3847 | .3355 | .2670 | .1977 | .1373 | .0896 | .0548 | .0312 |
| | 2 | .0515 | .1488 | .2376 | .2936 | .3115 | .2965 | .2587 | .2090 | .1569 | .1094 |
| | 3 | .0054 | .0331 | .0839 | .1468 | .2076 | .2541 | .2786 | .2787 | .2568 | .2188 |
| | 4 | .0004 | .0046 | .0185 | .0459 | .0865 | .1361 | .1875 | .2322 | .2627 | .2734 |
| | 5 | .0000 | .0004 | .0026 | .0092 | .0231 | .0467 | .0808 | .1239 | .1719 | .2188 |
| | 6 | .0000 | .0000 | .0002 | .0011 | .0038 | .0100 | .0217 | .0413 | .0703 | .1094 |
| | 7 | .0000 | .0000 | .0000 | .0001 | .0004 | .0012 | .0033 | .0079 | .0164 | .0312 |
| | 8 | .0000 | .0000 | .0000 | .0000 | .0000 | .0001 | .0002 | .0007 | .0017 | .0039 |

Linear interpolations with respect to $\theta$ will in general be accurate at most to two decimal places.

Table A.5 (continued)

**INDIVIDUAL TERMS, BINOMIAL DISTRIBUTION**

| | | | | | | θ | | | | | |
|---|---|---|---|---|---|---|---|---|---|---|---|
| *n* | *x* | .05 | .10 | .15 | .20 | .25 | .30 | .35 | .40 | .45 | .50 |
| 9 | 0 | .6302 | .3874 | .2316 | .1342 | .0751 | .0404 | .0207 | .0101 | .0046 | .0020 |
| | 1 | .2985 | .3874 | .3679 | .3020 | .2253 | .1556 | .1004 | .0605 | .0339 | .0176 |
| | 2 | .0629 | .1722 | .2597 | .3020 | .3003 | .2668 | .2162 | .1612 | .1110 | .0703 |
| | 3 | .0077 | .0446 | .1069 | .1762 | .2336 | .2668 | .2716 | .2508 | .2119 | .1641 |
| | 4 | .0006 | .0074 | .0283 | .0661 | .1168 | .1715 | .2194 | .2508 | .2600 | .2461 |
| | 5 | .0000 | .0008 | .0050 | .0165 | .0389 | .0735 | .1181 | .1672 | .2128 | .2461 |
| | 6 | .0000 | .0001 | .0006 | .0028 | .0087 | .0210 | .0424 | .0743 | .1160 | .1641 |
| | 7 | .0000 | .0000 | .0000 | .0003 | .0012 | .0039 | .0098 | .0212 | .0407 | .0703 |
| | 8 | .0000 | .0000 | .0000 | .0000 | .0001 | .0004 | .0013 | .0035 | .0083 | .0176 |
| | 9 | .0000 | .0000 | .0000 | .0000 | .0000 | .0000 | .0001 | .0003 | .0008 | .0020 |
| 10 | 0 | .5987 | .3487 | .1969 | .1074 | .0563 | .0282 | .0135 | .0060 | .0025 | .0010 |
| | 1 | .3151 | .3874 | .3474 | .2684 | .1877 | .1211 | .0725 | .0403 | .0207 | .0098 |
| | 2 | .0746 | .1937 | .2759 | .3020 | .2816 | .2335 | .1757 | .1209 | .0763 | .0439 |
| | 3 | .0105 | .0574 | .1298 | .2013 | .2503 | .2668 | .2522 | .2150 | .1665 | .1172 |
| | 4 | .0010 | .0112 | .0401 | .0881 | .1460 | .2001 | .2377 | .2508 | .2384 | .2051 |
| | 5 | .0001 | .0015 | .0085 | .0264 | .0584 | .1029 | .1536 | .2007 | .2340 | .2461 |
| | 6 | .0000 | .0001 | .0012 | .0055 | .0162 | .0368 | .0689 | .1115 | .1596 | .2051 |
| | 7 | .0000 | .0000 | .0001 | .0008 | .0031 | .0090 | .0212 | .0425 | .0746 | .1172 |
| | 8 | .0000 | .0000 | .0000 | .0001 | .0004 | .0014 | .0043 | .0106 | .0229 | .0439 |
| | 9 | .0000 | .0000 | .0000 | .0000 | .0000 | .0001 | .0005 | .0016 | .0042 | .0098 |
| | 10 | .0000 | .0000 | .0000 | .0000 | .0000 | .0000 | .0000 | .0001 | .0003 | .0010 |
| 11 | 0 | .5688 | .3138 | .1673 | .0859 | .0422 | .0198 | .0088 | .0036 | .0014 | .0004 |
| | 1 | .3293 | .3835 | .3248 | .2362 | .1549 | .0932 | .0518 | .0266 | .0125 | .0055 |
| | 2 | .0867 | .2131 | .2866 | .2953 | .2581 | .1998 | .1395 | .0887 | .0513 | .0269 |
| | 3 | .0137 | .0710 | .1517 | .2215 | .2581 | .2568 | .2254 | .1774 | .1259 | .0806 |
| | 4 | .0014 | .0158 | .0536 | .1107 | .1721 | .2201 | .2428 | .2365 | .2060 | .1611 |
| | 5 | .0001 | .0025 | .0132 | .0388 | .0803 | .1321 | .1830 | .2207 | .2360 | .2256 |
| | 6 | .0000 | .0003 | .0023 | .0097 | .0268 | .0566 | .0985 | .1471 | .1931 | .2256 |
| | 7 | .0000 | .0000 | .0003 | .0017 | .0064 | .0173 | .0379 | .0701 | .1128 | .1611 |
| | 8 | .0000 | .0000 | .0000 | .0002 | .0011 | .0037 | .0102 | .0234 | .0462 | .0806 |
| | 9 | .0000 | .0000 | .0000 | .0000 | .0001 | .0005 | .0018 | .0052 | .0126 | .0269 |
| | 10 | .0000 | .0000 | .0000 | .0000 | .0000 | .0000 | .0002 | .0007 | .0021 | .0054 |
| | 11 | .0000 | .0000 | .0000 | .0000 | .0000 | .0000 | .0000 | .0000 | .0002 | .0005 |
| 12 | 0 | .5404 | .2824 | .1422 | .0687 | .0317 | .0138 | .0057 | .0022 | .0008 | .0002 |
| | 1 | .3413 | .3766 | .3012 | .2062 | .1267 | .0712 | .0368 | .0174 | .0075 | .0029 |
| | 2 | .0988 | .2301 | .2924 | .2835 | .2323 | .1678 | .1088 | .0639 | .0339 | .0161 |
| | 3 | .0173 | .0852 | .1720 | .2362 | .2581 | .2397 | .1954 | .1419 | .0923 | .0537 |
| | 4 | .0021 | .0213 | .0683 | .1329 | .1936 | .2311 | .2367 | .2128 | .1700 | .1208 |
| | 5 | .0002 | .0038 | .0193 | .0532 | .1032 | .1585 | .2039 | .2270 | .2225 | .1934 |
| | 6 | .0000 | .0005 | .0040 | .0155 | .0401 | .0792 | .1281 | .1766 | .2124 | .2256 |
| | 7 | .0000 | .0000 | .0006 | .0033 | .0115 | .0291 | .0591 | .1009 | .1489 | .1934 |
| | 8 | .0000 | .0000 | .0001 | .0005 | .0024 | .0078 | .0199 | .0420 | .0762 | .1208 |
| | 9 | .0000 | .0000 | .0000 | .0001 | .0004 | .0015 | .0048 | .0125 | .0277 | .0537 |
| | 10 | .0000 | .0000 | .0000 | .0000 | .0000 | .0002 | .0008 | .0025 | .0068 | .0161 |
| | 11 | .0000 | .0000 | .0000 | .0000 | .0000 | .0000 | .0001 | .0003 | .0010 | .0029 |
| | 12 | .0000 | .0000 | .0000 | .0000 | .0000 | .0000 | .0000 | .0000 | .0001 | .0002 |

Table A.5 (continued)

**INDIVIDUAL TERMS, BINOMIAL DISTRIBUTION**

| | | | | | | $\theta$ | | | | | |
|---|---|---|---|---|---|---|---|---|---|---|---|
| $n$ | $x$ | .05 | .10 | .15 | .20 | .25 | .30 | .35 | .40 | .45 | .50 |
| 13 | 0 | .5133 | .2542 | .1209 | .0550 | .0238 | .0097 | .0037 | .0013 | .0004 | .0001 |
| | 1 | .3512 | .3672 | .2774 | .1787 | .1029 | .0540 | .0259 | .0113 | .0045 | .0016 |
| | 2 | .1109 | .2448 | .2937 | .2680 | .2059 | .1388 | .0836 | .0453 | .0220 | .0095 |
| | 3 | .0214 | .0997 | .1900 | .2457 | .2517 | .2181 | .1651 | .1107 | .0660 | .0349 |
| | 4 | .0028 | .0277 | .0838 | .1535 | .2097 | .2337 | .2222 | .1845 | .1350 | .0873 |
| | 5 | .0003 | .0055 | .0266 | .0691 | .1258 | .1803 | .2154 | .2214 | .1989 | .1571 |
| | 6 | .0000 | .0008 | .0063 | .0230 | .0559 | .1030 | .1546 | .1968 | .2169 | .2095 |
| | 7 | .0000 | .0001 | .0011 | .0058 | .0186 | .0442 | .0833 | .1312 | .1775 | .2095 |
| | 8 | .0000 | .0000 | .0001 | .0011 | .0047 | .0142 | .0336 | .0656 | .1089 | .1571 |
| | 9 | .0000 | .0000 | .0000 | .0001 | .0009 | .0034 | .0101 | .0243 | .0495 | .0873 |
| | 10 | .0000 | .0000 | .0000 | .0000 | .0001 | .0006 | .0022 | .0065 | .0162 | .0349 |
| | 11 | .0000 | .0000 | .0000 | .0000 | .0000 | .0001 | .0003 | .0012 | .0036 | .0095 |
| | 12 | .0000 | .0000 | .0000 | .0000 | .0000 | .0000 | .0000 | .0001 | .0005 | .0016 |
| | 13 | .0000 | .0000 | .0000 | .0000 | .0000 | .0000 | .0000 | .0000 | .0000 | .0001 |
| 14 | 0 | .4877 | .2288 | .1028 | .0440 | .0178 | .0068 | .0024 | .0008 | .0002 | .0001 |
| | 1 | .3593 | .3559 | .2539 | .1539 | .0832 | .0407 | .0181 | .0073 | .0027 | .0009 |
| | 2 | .1229 | .2570 | .2912 | .2501 | .1802 | .1134 | .0634 | .0317 | .0141 | .0056 |
| | 3 | .0259 | .1142 | .2056 | .2501 | .2402 | .1943 | .1366 | .0845 | .0462 | .0222 |
| | 4 | .0037 | .0349 | .0998 | .1720 | .2202 | .2290 | .2022 | .1549 | .1040 | .0611 |
| | 5 | .0004 | .0078 | .0352 | .0860 | .1468 | .1963 | .2178 | .2066 | .1701 | .1222 |
| | 6 | .0000 | .0013 | .0093 | .0322 | .0734 | .1262 | .1759 | .2066 | .2088 | .1833 |
| | 7 | .0000 | .0002 | .0019 | .0092 | .0280 | .0618 | .1082 | .1574 | .1952 | .2095 |
| | 8 | .0000 | .0000 | .0003 | .0020 | .0082 | .0232 | .0510 | .0918 | .1398 | .1833 |
| | 9 | .0000 | .0000 | .0000 | .0003 | .0018 | .0066 | .0183 | .0408 | .0762 | .1222 |
| | 10 | .0000 | .0000 | .0000 | .0000 | .0003 | .0014 | .0049 | .0136 | .0312 | .0611 |
| | 11 | .0000 | .0000 | .0000 | .0000 | .0000 | .0002 | .0010 | .0033 | .0093 | .0222 |
| | 12 | .0000 | .0000 | .0000 | .0000 | .0000 | .0000 | .0001 | .0005 | .0019 | .0056 |
| | 13 | .0000 | .0000 | .0000 | .0000 | .0000 | .0000 | .0000 | .0001 | .0002 | .0009 |
| | 14 | .0000 | .0000 | .0000 | .0000 | .0000 | .0000 | .0000 | .0000 | .0000 | .0001 |
| 15 | 0 | .4633 | .2059 | .0874 | .0352 | .0134 | .0047 | .0016 | .0005 | .0001 | .0000 |
| | 1 | .3658 | .3432 | .2312 | .1319 | .0668 | .0305 | .0126 | .0047 | .0016 | .0005 |
| | 2 | .1348 | .2669 | .2856 | .2309 | .1559 | .0916 | .0476 | .0219 | .0090 | .0032 |
| | 3 | .0307 | .1285 | .2184 | .2501 | .2252 | .1700 | .1110 | .0634 | .0318 | .0139 |
| | 4 | .0049 | .0428 | .1156 | .1876 | .2252 | .2186 | .1792 | .1268 | .0780 | .0417 |
| | 5 | .0006 | .0105 | .0449 | .1032 | .1651 | .2061 | .2123 | .1859 | .1404 | .0916 |
| | 6 | .0000 | .0019 | .0132 | .0430 | .0917 | .1472 | .1906 | .2066 | .1914 | .1527 |
| | 7 | .0000 | .0003 | .0030 | .0138 | .0393 | .0811 | .1319 | .1771 | .2013 | .1964 |
| | 8 | .0000 | .0000 | .0005 | .0035 | .0131 | .0348 | .0710 | .1181 | .1647 | .1964 |
| | 9 | .0000 | .0000 | .0001 | .0007 | .0034 | .0116 | .0298 | .0612 | .1048 | .1527 |
| | 10 | .0000 | .0000 | .0000 | .0001 | .0007 | .0030 | .0096 | .0245 | .0515 | .0916 |
| | 11 | .0000 | .0000 | .0000 | .0000 | .0001 | .0006 | .0024 | .0074 | .0191 | .0417 |
| | 12 | .0000 | .0000 | .0000 | .0000 | .0000 | .0001 | .0004 | .0016 | .0052 | .0139 |
| | 13 | .0000 | .0000 | .0000 | .0000 | .0000 | .0000 | .0001 | .0003 | .0010 | .0032 |
| | 14 | .0000 | .0000 | .0000 | .0000 | .0000 | .0000 | .0000 | .0000 | .0001 | .0005 |
| | 15 | .0000 | .0000 | .0000 | .0000 | .0000 | .0000 | .0000 | .0000 | .0000 | .0000 |

Table A.5 (continued)

**INDIVIDUAL TERMS, BINOMIAL DISTRIBUTION**

| | | | | | | $\theta$ | | | | | |
|---|---|---|---|---|---|---|---|---|---|---|---|
| $n$ | $x$ | .05 | .10 | .15 | .20 | .25 | .30 | .35 | .40 | .45 | .50 |
| 16 | 0 | .4401 | .1853 | .0743 | .0281 | .0100 | .0033 | .0010 | .0003 | .0001 | .0000 |
| | 1 | .3706 | .3294 | .2097 | .1126 | .0535 | .0228 | .0087 | .0030 | .0009 | .0002 |
| | 2 | .1463 | .2745 | .2775 | .2111 | .1336 | .0732 | .0353 | .0150 | .0056 | .0018 |
| | 3 | .0359 | .1423 | .2285 | .2463 | .2079 | .1465 | .0888 | .0468 | .0215 | .0085 |
| | 4 | .0061 | .0514 | .1311 | .2001 | .2252 | .2040 | .1553 | .1014 | .0572 | .0278 |
| | 5 | .0008 | .0137 | .0555 | .1201 | .1802 | .2099 | .2008 | .1623 | .1123 | .0667 |
| | 6 | .0001 | .0028 | .0180 | .0550 | .1101 | .1649 | .1982 | .1983 | .1684 | .1222 |
| | 7 | .0000 | .0004 | .0045 | .0197 | .0524 | .1010 | .1524 | .1889 | .1969 | .1746 |
| | 8 | .0000 | .0001 | .0009 | .0055 | .0197 | .0487 | .0923 | .1417 | .1812 | .1964 |
| | 9 | .0000 | .0000 | .0001 | .0012 | .0058 | .0185 | .0442 | .0840 | .1318 | .1746 |
| | 10 | .0000 | .0000 | .0000 | .0002 | .0014 | .0056 | .0167 | .0392 | .0755 | .1222 |
| | 11 | .0000 | .0000 | .0000 | .0000 | .0002 | .0013 | .0049 | .0142 | .0337 | .0667 |
| | 12 | .0000 | .0000 | .0000 | .0000 | .0000 | .0002 | .0011 | .0040 | .0115 | .0278 |
| | 13 | .0000 | .0000 | .0000 | .0000 | .0000 | .0000 | .0002 | .0008 | .0029 | .0085 |
| | 14 | .0000 | .0000 | .0000 | .0000 | .0000 | .0000 | .0000 | .0001 | .0005 | .0018 |
| | 15 | .0000 | .0000 | .0000 | .0000 | .0000 | .0000 | .0000 | .0000 | .0001 | .0002 |
| | 16 | .0000 | .0000 | .0000 | .0000 | .0000 | .0000 | .0000 | .0000 | .0000 | .0000 |
| 17 | 0 | .4181 | .1668 | .0631 | .0225 | .0075 | .0023 | .0007 | .0002 | .0000 | .0000 |
| | 1 | .3741 | .3150 | .1893 | .0957 | .0426 | .0169 | .0060 | .0019 | .0005 | .0001 |
| | 2 | .1575 | .2800 | .2673 | .1914 | .1136 | .0581 | .0260 | .0102 | .0035 | .0010 |
| | 3 | .0415 | .1556 | .2359 | .2393 | .1893 | .1245 | .0701 | .0341 | .0144 | .0052 |
| | 4 | .9076 | .0605 | .1457 | .2093 | .2209 | .1868 | .1320 | .0796 | .0411 | .0182 |
| | 5 | .0010 | .0175 | .0668 | .1361 | .1914 | .2081 | .1849 | .1379 | .0875 | .0472 |
| | 6 | .0001 | .0039 | .0236 | .0680 | .1276 | .1784 | .1991 | .1839 | .1432 | .0944 |
| | 7 | .0000 | .0007 | .0065 | .0267 | .0668 | .1201 | .1685 | .1927 | .1841 | .1484 |
| | 8 | .0000 | .0001 | .0014 | .0084 | .0279 | .0644 | .1134 | .1606 | .1883 | .1855 |
| | 9 | .0000 | .0000 | .0003 | .0021 | .0093 | .0276 | .0611 | .1070 | .1540 | .1855 |
| | 10 | .0000 | .0000 | .0000 | .0004 | .0025 | .0095 | .0263 | .0571 | .1008 | .1484 |
| | 11 | .0000 | .0000 | .0000 | .0001 | .0005 | .0026 | .0090 | .0242 | .0525 | .0944 |
| | 12 | .0000 | .0000 | .0000 | .0000 | .0001 | .0006 | .0024 | .0081 | .0215 | .0472 |
| | 13 | .0000 | .0000 | .0000 | .0000 | .0000 | .0001 | .0005 | .0021 | .0068 | .0182 |
| | 14 | .0000 | .0000 | .0000 | .0000 | .0000 | .0000 | .0001 | .0004 | .0016 | .0052 |
| | 15 | .0000 | .0000 | .0000 | .0000 | .0000 | .0000 | .0000 | .0001 | .0003 | .0010 |
| | 16 | .0000 | .0000 | .0000 | .0000 | .0000 | .0000 | .0000 | .0000 | .0000 | .0001 |
| | 17 | .0000 | .0000 | .0000 | .0000 | .0000 | .0000 | .0000 | .0000 | .0000 | .0000 |
| 18 | 0 | .3972 | .1501 | .0536 | .0180 | .0056 | .0016 | .0004 | .0001 | .0000 | .0000 |
| | 1 | .3763 | .3002 | .1704 | .0811 | .0338 | .0126 | .0042 | .0012 | .0003 | .0001 |
| | 2 | .1683 | .2835 | .2556 | .1723 | .0958 | .0458 | .0190 | .0069 | .0022 | .0006 |
| | 3 | .0473 | .1680 | .2406 | .2297 | .1704 | .1046 | .0547 | .0246 | .0095 | .0031 |
| | 4 | .0093 | .0700 | .1592 | .2153 | .2130 | .1681 | .1104 | .0614 | .0291 | .0117 |
| | 5 | .0014 | .0218 | .0787 | .1507 | .1988 | .2017 | .1664 | .1146 | .0666 | .0327 |
| | 6 | .0002 | .0052 | .0301 | .0816 | .1436 | .1873 | .1941 | .1655 | .1181 | .0708 |
| | 7 | .0000 | .0010 | .0091 | .0350 | .0820 | .1376 | .1792 | .1892 | .1657 | .1214 |
| | 8 | .0000 | .0002 | .0022 | .0120 | .0376 | .0811 | .1327 | .1734 | .1864 | .1669 |
| | 9 | .0000 | .0000 | .0004 | .0033 | .0139 | .0386 | .0794 | .1284 | .1694 | .1855 |
| | 10 | .0000 | .0000 | .0001 | .0008 | .0042 | .0149 | .0385 | .0771 | .1248 | .1669 |
| | 11 | .0000 | .0000 | .0000 | .0001 | .0010 | .0046 | .0151 | .0374 | .0742 | .1214 |

Table A.5 (continued)

**INDIVIDUAL TERMS, BINOMIAL DISTRIBUTION**

| $n$ | $x$ | $\theta$ .05 | .10 | .15 | .20 | .25 | .30 | .35 | .40 | .45 | .50 |
|---|---|---|---|---|---|---|---|---|---|---|---|
| 18 | 12 | .0000 | .0000 | .0000 | .0000 | .0002 | .0012 | .0047 | .0145 | .0354 | .0708 |
| | 13 | .0000 | .0000 | .0000 | .0000 | .0000 | .0002 | .0012 | .0045 | .0134 | .0327 |
| | 14 | .0000 | .0000 | .0000 | .0000 | .0000 | .0000 | .0002 | .0011 | .0039 | .0117 |
| | 15 | .0000 | .0000 | .0000 | .0000 | .0000 | .0000 | .0000 | .0002 | .0009 | .0031 |
| | 16 | .0000 | .0000 | .0000 | .0000 | .0000 | .0000 | .0000 | .0000 | .0001 | .0006 |
| | 17 | .0000 | .0000 | .0000 | .0000 | .0000 | .0000 | .0000 | .0000 | .0000 | .0001 |
| | 18 | .0000 | .0000 | .0000 | .0000 | .0000 | .0000 | .0000 | .0000 | .0000 | .0000 |
| 19 | 0 | .3774 | .1351 | .0456 | .0144 | .0042 | .0011 | .0003 | .0001 | .0000 | .0000 |
| | 1 | .3774 | .2852 | .1529 | .0685 | .0268 | .0093 | .0029 | .0008 | .0002 | .0000 |
| | 2 | .1787 | .2852 | .2428 | .1540 | .0803 | .0358 | .0138 | .0046 | .0013 | .0003 |
| | 3 | .0533 | .1796 | .2428 | .2182 | .1517 | .0869 | .0422 | .0175 | .0062 | .0018 |
| | 4 | .0112 | .0798 | .1714 | .2182 | .2023 | .1491 | .0909 | .0467 | .0203 | .0074 |
| | 5 | .0018 | .0266 | .0907 | .1636 | .2023 | .1916 | .1468 | .0933 | .0497 | .0222 |
| | 6 | .0002 | .0069 | .0374 | .0955 | .1574 | .1916 | .1844 | .1451 | .0949 | .0518 |
| | 7 | .0000 | .0014 | .0122 | .0443 | .0974 | .1525 | .1844 | .1797 | .1443 | .0961 |
| | 8 | .0000 | .0002 | .0032 | .0166 | .0487 | .0981 | .1489 | .1797 | .1771 | .1442 |
| | 9 | .0000 | .0000 | .0007 | .0051 | .0198 | .0514 | .0980 | .1464 | .1771 | .1762 |
| | 10 | .0000 | .0000 | .0001 | .0013 | .0066 | .0220 | .0528 | .0976 | .1449 | .1762 |
| | 11 | .0000 | .0000 | .0000 | .0003 | .0018 | .0077 | .0233 | .0532 | .0970 | .1442 |
| | 12 | .0000 | .0000 | .0000 | .0000 | .0004 | .0022 | .0083 | .0237 | .0529 | .0961 |
| | 13 | .0000 | .0000 | .0000 | .0000 | .0001 | .0005 | .0024 | .0085 | .0233 | .0518 |
| | 14 | .0000 | .0000 | .0000 | .0000 | .0000 | .0001 | .0006 | .0024 | .0082 | .0222 |
| | 15 | .0000 | .0000 | .0000 | .0000 | .0000 | .0000 | .0001 | .0005 | .0022 | .0074 |
| | 16 | .0000 | .0000 | .0000 | .0000 | .0000 | .0000 | .0000 | .0001 | .0005 | .0018 |
| | 17 | .0000 | .0000 | .0000 | .0000 | .0000 | .0000 | .0000 | .0000 | .0001 | .0003 |
| | 18 | .0000 | .0000 | .0000 | .0000 | .0000 | .0000 | .0000 | .0000 | .0000 | .0000 |
| | 19 | .0000 | .0000 | .0000 | .0000 | .0000 | .0000 | .0000 | .0000 | .0000 | .0000 |
| 20 | 0 | .3585 | .1216 | .0388 | .0115 | .0032 | .0008 | .0002 | .0000 | .0000 | .0000 |
| | 1 | .3774 | .2702 | .1368 | .0576 | .0211 | .0068 | .0020 | .0005 | .0001 | .0000 |
| | 2 | .1887 | .2852 | .2293 | .1369 | .0669 | .0278 | .0100 | .0031 | .0008 | .0002 |
| | 3 | .0596 | .1901 | .2428 | .2054 | .1339 | .0716 | .0323 | .0123 | .0040 | .0011 |
| | 4 | .0133 | .0898 | .1821 | .2182 | .1897 | .1304 | .0738 | .0350 | .0139 | .0046 |
| | 5 | .0022 | .0319 | .1028 | .1746 | .2023 | .1789 | .1272 | .0746 | .0365 | .0148 |
| | 6 | .0003 | .0089 | .0454 | .1091 | .1686 | .1916 | .1712 | .1244 | .0746 | .0370 |
| | 7 | .0000 | .0020 | .0160 | .0545 | .1124 | .1643 | .1844 | .1659 | .1221 | .0739 |
| | 8 | .0000 | .0004 | .0046 | .0222 | .0609 | .1144 | .1614 | .1797 | .1623 | .1201 |
| | 9 | .0000 | .0001 | .0011 | .0074 | .0271 | .0654 | .1158 | .1597 | .1771 | .1602 |
| | 10 | .0000 | .0000 | .0002 | .0020 | .0099 | .0308 | .0686 | .1171 | .1593 | .1762 |
| | 11 | .0000 | .0000 | .0000 | .0005 | .0030 | .0120 | .0336 | .0710 | .1185 | .1602 |
| | 12 | .0000 | .0000 | .0000 | .0001 | .0008 | .0039 | .0136 | .0355 | .0727 | .1201 |
| | 13 | .0000 | .0000 | .0000 | .0000 | .0002 | .0010 | .0045 | .0146 | .0366 | .0739 |
| | 14 | .0000 | .0000 | .0000 | .0000 | .0000 | .0002 | .0012 | .0049 | .0150 | .0370 |
| | 15 | .0000 | .0000 | .0000 | .0000 | .0000 | .0000 | .0003 | .0013 | .0049 | .0148 |
| | 16 | .0000 | .0000 | .0000 | .0000 | .0000 | .0000 | .0000 | .0003 | .0013 | .0046 |
| | 17 | .0000 | .0000 | .0000 | .0000 | .0000 | .0000 | .0000 | .0000 | .0002 | .0011 |
| | 18 | .0000 | .0000 | .0000 | .0000 | .0000 | .0000 | .0000 | .0000 | .0000 | .0002 |
| | 19 | .0000 | .0000 | .0000 | .0000 | .0000 | .0000 | .0000 | .0000 | .0000 | .0000 |
| | 20 | .0000 | .0000 | .0000 | .0000 | .0000 | .0000 | .0000 | .0000 | .0000 | .0000 |

Table A.6 Random Numbers (The first 5 decimals).

| Line/Col. | (1) | (2) | (3) | (4) | (5) | (6) | (7) | (8) | (9) | (10) | (11) | (12) | (13) | (14) |
|---|---|---|---|---|---|---|---|---|---|---|---|---|---|---|
| 1 | 10480 | 15011 | 01536 | 02011 | 81647 | 91646 | 69179 | 14194 | 62590 | 36207 | 20969 | 99570 | 91291 | 90700 |
| 2 | 22368 | 46573 | 25595 | 85393 | 30995 | 89198 | 27982 | 53402 | 93965 | 34095 | 52666 | 19174 | 39615 | 99505 |
| 3 | 24130 | 48360 | 22527 | 97265 | 76393 | 64809 | 15179 | 24830 | 49340 | 32081 | 30680 | 19655 | 63348 | 58629 |
| 4 | 42167 | 93093 | 06243 | 61680 | 07856 | 16376 | 39440 | 53537 | 71341 | 57004 | 00849 | 74917 | 97758 | 16379 |
| 5 | 37570 | 39975 | 81837 | 16656 | 06121 | 91782 | 60468 | 81305 | 49684 | 60672 | 14110 | 06927 | 01263 | 54613 |
| 6 | 77921 | 06907 | 11008 | 42751 | 27756 | 53498 | 18602 | 70659 | 90655 | 15053 | 21916 | 81825 | 44394 | 42880 |
| 7 | 99562 | 72905 | 56420 | 69994 | 98872 | 31016 | 71194 | 18738 | 44013 | 48840 | 63213 | 21069 | 10634 | 12952 |
| 8 | 96301 | 91977 | 05463 | 07972 | 18876 | 20922 | 94595 | 56869 | 69014 | 60045 | 18425 | 84903 | 42508 | 32307 |
| 9 | 89579 | 14342 | 63661 | 10281 | 17453 | 18103 | 57740 | 84378 | 25331 | 12566 | 58678 | 44947 | 05585 | 56941 |
| 10 | 85475 | 36857 | 43342 | 53988 | 53060 | 59533 | 38867 | 62300 | 08158 | 17983 | 16439 | 11458 | 18593 | 64952 |
| 11 | 28918 | 69578 | 88231 | 33276 | 70997 | 79936 | 56865 | 05859 | 90106 | 31595 | 01547 | 85590 | 91610 | 78188 |
| 12 | 63553 | 40961 | 48235 | 03427 | 49626 | 69445 | 18663 | 72695 | 52180 | 20847 | 12234 | 90511 | 33703 | 90322 |
| 13 | 09429 | 93969 | 52636 | 92737 | 88974 | 33488 | 36320 | 17617 | 30015 | 08272 | 84115 | 27156 | 30613 | 74952 |
| 14 | 10365 | 61129 | 87529 | 85689 | 48237 | 52267 | 67689 | 93394 | 01511 | 26358 | 85104 | 20285 | 29975 | 89868 |
| 15 | 07119 | 97336 | 71048 | 08178 | 77233 | 13916 | 47564 | 81056 | 97735 | 85977 | 29372 | 74461 | 28551 | 90707 |
| 16 | 51085 | 12765 | 51821 | 51259 | 77452 | 16308 | 60756 | 92144 | 49442 | 53900 | 70960 | 63990 | 75601 | 40719 |
| 17 | 02368 | 21382 | 52404 | 60268 | 89368 | 19885 | 55322 | 44819 | 01188 | 65255 | 64835 | 44919 | 05944 | 55157 |
| 18 | 01011 | 54092 | 33362 | 94904 | 31273 | 04146 | 18594 | 29852 | 71585 | 85030 | 51132 | 01915 | 92747 | 64951 |
| 19 | 52162 | 53916 | 46369 | 58586 | 23216 | 14513 | 83149 | 98736 | 23495 | 64350 | 94738 | 17752 | 35156 | 35749 |
| 20 | 07056 | 97628 | 33787 | 09998 | 42698 | 06691 | 76988 | 13602 | 51851 | 46104 | 88916 | 19509 | 25625 | 58104 |
| 21 | 48663 | 91245 | 85828 | 14346 | 09172 | 30168 | 90229 | 04734 | 59193 | 22178 | 30421 | 61666 | 99904 | 32812 |
| 22 | 54164 | 58492 | 22421 | 74103 | 47070 | 25306 | 76468 | 26384 | 58151 | 06646 | 21524 | 15227 | 96909 | 44592 |
| 23 | 32639 | 32363 | 05597 | 24200 | 13363 | 38005 | 94342 | 28728 | 35806 | 06912 | 17012 | 64161 | 18296 | 22851 |
| 24 | 29334 | 27001 | 87637 | 87308 | 58731 | 00256 | 45834 | 15398 | 46557 | 41135 | 10367 | 07684 | 36188 | 18510 |
| 25 | 02488 | 33062 | 28834 | 07351 | 19731 | 92420 | 60952 | 61280 | 50001 | 67658 | 32586 | 86679 | 50720 | 94953 |
| 26 | 81525 | 72295 | 04839 | 96423 | 24878 | 82651 | 66566 | 14778 | 76797 | 14780 | 13300 | 87074 | 79666 | 95725 |
| 27 | 29676 | 20591 | 68086 | 26432 | 46901 | 20849 | 89768 | 81536 | 86645 | 12659 | 92259 | 57102 | 80428 | 25280 |
| 28 | 00742 | 57392 | 39064 | 66432 | 84673 | 40027 | 32832 | 61362 | 98947 | 96067 | 64760 | 64584 | 96096 | 98253 |
| 29 | 05366 | 04213 | 25669 | 26422 | 44407 | 44048 | 37937 | 63904 | 45766 | 66134 | 75470 | 66520 | 34693 | 90449 |
| 30 | 91921 | 26418 | 64117 | 94305 | 26766 | 25940 | 39972 | 22209 | 71500 | 64568 | 91402 | 42416 | 07844 | 69618 |
| 31 | 00582 | 04711 | 87917 | 77341 | 42206 | 35126 | 74087 | 99547 | 81817 | 42607 | 43808 | 76655 | 62028 | 76630 |
| 32 | 00725 | 69884 | 62797 | 56170 | 86324 | 88072 | 76222 | 36086 | 84637 | 93161 | 76038 | 65855 | 77919 | 88006 |
| 33 | 69011 | 65797 | 95876 | 55293 | 18988 | 27354 | 26575 | 08625 | 40801 | 59920 | 29841 | 80150 | 12777 | 48501 |
| 34 | 25976 | 57948 | 29888 | 88604 | 67917 | 48708 | 18912 | 82271 | 65424 | 69774 | 33611 | 54262 | 85963 | 03547 |
| 35 | 09763 | 83473 | 73577 | 12908 | 30883 | 18317 | 28290 | 35797 | 05998 | 41688 | 34952 | 37888 | 38917 | 88050 |
| 36 | 91567 | 42595 | 27958 | 30134 | 04024 | 86385 | 29880 | 99730 | 55536 | 84855 | 29080 | 09250 | 79656 | 73211 |
| 37 | 17955 | 56349 | 90999 | 49127 | 20044 | 59931 | 06115 | 20542 | 18059 | 02008 | 73708 | 83517 | 36103 | 42791 |
| 38 | 46503 | 18584 | 18845 | 49618 | 02304 | 51038 | 20655 | 58727 | 28168 | 15475 | 56942 | 53389 | 20562 | 87338 |
| 39 | 92157 | 89634 | 94824 | 78171 | 84610 | 82834 | 09922 | 25417 | 44137 | 48413 | 25555 | 21246 | 35509 | 20468 |
| 40 | 14577 | 62765 | 35605 | 81263 | 39667 | 47358 | 56873 | 56307 | 61607 | 49518 | 89656 | 20103 | 77490 | 18062 |
| 41 | 98427 | 07523 | 33362 | 64270 | 01638 | 92477 | 66969 | 98420 | 04880 | 45585 | 46565 | 04102 | 46880 | 45709 |
| 42 | 34914 | 63976 | 88720 | 82765 | 34476 | 17032 | 87589 | 40836 | 32427 | 70002 | 70663 | 88863 | 77775 | 69348 |
| 43 | 70060 | 28277 | 39475 | 46473 | 23219 | 53416 | 94970 | 25832 | 69975 | 94884 | 19661 | 72828 | 00102 | 66794 |
| 44 | 53976 | 54914 | 06990 | 67245 | 68350 | 82948 | 11398 | 42878 | 80287 | 88267 | 47363 | 46634 | 06541 | 97809 |
| 45 | 76072 | 29515 | 40980 | 07391 | 58745 | 25774 | 22987 | 80059 | 39911 | 96189 | 41151 | 14222 | 60697 | 59583 |
| 46 | 90725 | 52210 | 83974 | 29992 | 65831 | 38857 | 50490 | 83765 | 55657 | 14361 | 31720 | 57375 | 56228 | 41546 |
| 47 | 64364 | 67412 | 33339 | 31926 | 14883 | 24413 | 59744 | 92351 | 97473 | 89286 | 35931 | 04110 | 23726 | 51900 |
| 48 | 08962 | 00358 | 31662 | 25388 | 61642 | 34072 | 81249 | 35648 | 56891 | 69352 | 48373 | 45578 | 78547 | 81788 |
| 49 | 95012 | 68379 | 93526 | 70765 | 10593 | 04542 | 76463 | 54328 | 02349 | 17247 | 28865 | 14777 | 62730 | 92277 |
| 50 | 15664 | 10493 | 20492 | 38391 | 91132 | 21999 | 59516 | 81652 | 27195 | 48223 | 46751 | 22923 | 32261 | 85653 |

Table A.7 Standard Normal Deviates ($\mu = 0, \sigma = 1$).

| 01 | 02 | 03 | 04 | 05 | 06 | 07 | 08 | 09 | 10 |
|---|---|---|---|---|---|---|---|---|---|
| 0.464 | 0.137 | 2.455 | −0.323 | −0.068 | 0.296 | −0.288 | 1.298 | 0.241 | −0.957 |
| 0.060 | −2.526 | −0.531 | −0.194 | 0.543 | −1.558 | 0.187 | −1.190 | 0.022 | 0.525 |
| 1.486 | −0.354 | −0.634 | 0.697 | 0.926 | 1.375 | 0.785 | −0.963 | −0.853 | −1.865 |
| 1.022 | −0.472 | 1.279 | 3.521 | 0.571 | −1.851 | 0.194 | 1.192 | −0.501 | −0.273 |
| 1.394 | −0.555 | 0.046 | 0.321 | 2.945 | 1.974 | −0.258 | 0.412 | 0.439 | −0.035 |
| 0.906 | −0.513 | −0.525 | 0.595 | 0.881 | −0.934 | 1.579 | 0.161 | −1.885 | 0.371 |
| 1.179 | −1.055 | 0.007 | 0.769 | 0.971 | 0.712 | 1.090 | −0.631 | −0.255 | −0.702 |
| −1.501 | −0.488 | −0.162 | −0.136 | 1.033 | 0.203 | 0.448 | 0.748 | −0.423 | −0.432 |
| −0.690 | 0.756 | −1.618 | −0.345 | −0.511 | −2.051 | −0.457 | −0.218 | 0.857 | −0.465 |
| 1.372 | 0.225 | 0.378 | 0.761 | 0.181 | −0.736 | 0.960 | −1.530 | −0.260 | 0.120 |
| −0.482 | 1.678 | −0.057 | −1.229 | −0.486 | 0.856 | −0.491 | −1.983 | −2.830 | −0.238 |
| −1.376 | −0.150 | 1.356 | −0.561 | −0.256 | −0.212 | 0.219 | 0.779 | 0.953 | −0.869 |
| −1.010 | 0.598 | −0.918 | 1.598 | 0.065 | 0.415 | −0.169 | 0.313 | −0.973 | −1.016 |
| −0.005 | −0.899 | 0.012 | −0.725 | 1.147 | −0.121 | 1.096 | 0.481 | −1.691 | 0.417 |
| 1.393 | −1.163 | −0.911 | 1.231 | −0.199 | −0.246 | 1.239 | −2.574 | −0.558 | 0.056 |
| −1.787 | −0.261 | 1.237 | 1.046 | −0.508 | −1.630 | −0.146 | −0.392 | −0.627 | 0.561 |
| −0.105 | −0.357 | −1.384 | 0.360 | −0.992 | −0.116 | −1.698 | −2.832 | −1.108 | −2.357 |
| −1.339 | 1.827 | −0.959 | 0.424 | 0.969 | −1.141 | −1.041 | 0.362 | −1.726 | 1.956 |
| 1.041 | 0.535 | 0.731 | 1.377 | 0.983 | −1.330 | 1.620 | −1.040 | 0.524 | −0.281 |
| 0.279 | −2.056 | 0.717 | −0.873 | −1.096 | −1.396 | 1.047 | 0.089 | −0.573 | 0.932 |
| −1.805 | −2.008 | −1.633 | 0.542 | 0.250 | −0.166 | 0.032 | 0.079 | 0.471 | −1.029 |
| −1.186 | 1.180 | 1.114 | 0.882 | 1.265 | −0.202 | 0.151 | −0.376 | −0.310 | 0.479 |
| 0.658 | −1.141 | 1.151 | −1.210 | −0.927 | 0.425 | 0.290 | −0.902 | 0.610 | 2.709 |
| −0.439 | 0.358 | −1.939 | 0.891 | −0.227 | 0.602 | 0.873 | −0.437 | −0.220 | −0.057 |
| −1.399 | −0.230 | 0.385 | −0.649 | −0.577 | 0.237 | −0.289 | 0.513 | 0.738 | −0.300 |
| 0.199 | 0.208 | −1.083 | −0.219 | −0.291 | 1.221 | 1.119 | 0.004 | −2.015 | −0.594 |
| 0.159 | 0.272 | −0.313 | 0.084 | −2.828 | −0.439 | −0.792 | −1.275 | −0.623 | −1.047 |
| 2.273 | 0.606 | 0.606 | −0.747 | 0.247 | 1.291 | 0.063 | −1.793 | −0.699 | −1.347 |
| 0.041 | −0.307 | 0.121 | 0.790 | −0.584 | 0.541 | 0.484 | −0.986 | 0.481 | 0.996 |
| −1.132 | −2.098 | 0.921 | 0.145 | 0.446 | −1.661 | 1.045 | −1.363 | −0.586 | −1.023 |
| 0.768 | 0.079 | −1.473 | 0.034 | −2.127 | 0.665 | 0.084 | −0.880 | −0.579 | 0.551 |
| 0.375 | −1.658 | −0.851 | 0.234 | −0.656 | 0.340 | −0.086 | −0.158 | −0.120 | 0.418 |
| −0.513 | −0.344 | 0.210 | −0.735 | 1.041 | 0.008 | 0.427 | −0.831 | 0.191 | 0.074 |
| 0.292 | −0.521 | 1.266 | −1.206 | −0.899 | 0.110 | −0.528 | −0.813 | 0.071 | 0.524 |
| 1.026 | 2.990 | −0.574 | −0.491 | −1.114 | 1.297 | −1.433 | −1.345 | −3.001 | 0.479 |
| −1.334 | 1.278 | −0.568 | −0.109 | −0.515 | −0.566 | 2.923 | 0.500 | 0.359 | 0.326 |
| −0.287 | −0.144 | −0.254 | 0.574 | −0.451 | −1.181 | −1.190 | −0.318 | −0.094 | 1.114 |
| 0.161 | −0.886 | −0.921 | −0.509 | 1.410 | −0.518 | 0.192 | −0.432 | 1.501 | 1.068 |
| −1.346 | 0.193 | −1.202 | 0.394 | −1.045 | 0.843 | 0.942 | 1.045 | 0.031 | 0.772 |
| 1.250 | −0.199 | −0.288 | 1.810 | 1.378 | 0.584 | 1.216 | 0.733 | 0.402 | 0.226 |
| 0.630 | −0.537 | 0.782 | 0.060 | 0.499 | −0.431 | 1.705 | 1.164 | 0.884 | −0.298 |
| 0.375 | −1.941 | 0.247 | −0.491 | −0.665 | −0.135 | −0.145 | −0.498 | 0.457 | 1.064 |
| −1.420 | 0.489 | −1.711 | −1.186 | 0.754 | −0.732 | −0.066 | 1.006 | −0.798 | 0.162 |
| −0.151 | −0.243 | −0.430 | −0.762 | 0.298 | 1.049 | 1.810 | 2.885 | −0.768 | −0.129 |
| −0.309 | 0.531 | 0.416 | −1.541 | 1.456 | 2.040 | −0.124 | 0.196 | 0.023 | −1.204 |
| 0.424 | −0.444 | 0.593 | 0.993 | −0.106 | 0.116 | 0.484 | −1.272 | 1.066 | 1.097 |
| 0.593 | 0.658 | −1.127 | −1.407 | −1.579 | −1.616 | 1.458 | 1.262 | 0.736 | −0.916 |
| 0.862 | −0.885 | −0.142 | −0.504 | 0.532 | 1.381 | 0.022 | −0.281 | −0.342 | 1.222 |
| 0.235 | −0.628 | −0.023 | −0.463 | −0.899 | −0.394 | −0.538 | 1.707 | −0.188 | −1.153 |
| −0.853 | 0.402 | 0.777 | 0.833 | 0.410 | −0.349 | −1.094 | 0.580 | 1.395 | 1.298 |

# INDEX

## Answers to Selected Exercises

### Exercise 1

1.1 2, 3, 5
1.2 No, it can generate at most 7 different numbers.
1.3 See whether it is uniform.
1.4 (iii) 820, 984.6, 1293.2
1.5 1,045
1.6 No integer division is used for modulus operation.
1.8 $S = \{(i, j)\ i, j = 1, 2, 3, 4, 5, 6\}$ (a) 1/9 (b) 11/36 (c) 1/6 (d) 25/36
1.9 (a) 2/11 (b) 6/25
1.11 (a) 1/13,000 (b) 1/130
1.12 (a) 0.264 (b) 0.44

### Exercise 2

2.1 $\mu = 3.5, \sigma^2 = 2.9167$
2.5 (0.387, 0.368), (0.387, 0.368), (0.187, 0.184)
2.6 (a) 0.368 (b) $.981^{10} = 0.825$
2.7 0.0001
2.8 (i) 0.84 (ii) 0.7327 (iii) 0.617 (iv) 0.98 (v) 0.42
2.9 0.01
2.10 57
2.12 $1/\lambda$
2.13 $e^{-1.5}$
2.14 Pr{Waiting time longer than $T$} = Pr{No call in the interval $[0, T]$}
2.16 Ag2
2.17 average time: Ag1 = 11.75, Ag2 = 11.5
2.18 (i) 257 (ii) 132 (iii) 451 (iv) 196 (Your answer may differ some.)
2.19 (i) Acceptance-rejection (ii) Inverse CDF (ii) Acceptance-rejection

### Exercise 3

3.1 (0.12, 0.28)
3.2 664
3.3 No. Risk prob. = 0.3918
3.4 Yes. Risk prob. = 0.004
3.5 (a) risk = 0.058 (b) risk = 0.116

3.6 (a) $p$-value = 0.042 (b) $H_0: p_1 = p_2 = \ldots = p_{10} = 1/10, H_1$: One of them is not 1/10. The risk probability can be found by simulation. It is 0.37.

3.7 472

3.8 (i) 0.23 (ii) Use Poisson 0.271 (iii) Use normal 0.00718

3.9 (a) 0.676 (b) 0.080 (c) $n = 1281, c = 17$

**Exercise 4**

4.4 $\hat{\mu} = \bar{x}$, $\hat{\sigma}^2 = (n-1)S^2/n$

4.5 (i) No. (ii) $\bar{\alpha}$ is not a good measure of the mean direction.

4.6 Yes.

4.7 Yes.

4.8 They can be reduced to two parameters.

**Exercise 5**

5.2 $p = 0.03$

5.3 $321 \pm 7.84$

5.4 $0.6 \pm 0.085$

5.5 None is significant at 0.05 level.

5.7 Mean interval is robust against non-normality, but the variance interval is not.

5.8 $Q = 40.55$, $p < 0.001$

5.9 $Q = 22.5$, $p < 0.001$. The breakdowns are not random.

5.10 $Q = 21.63$, $p < 0.001$

5.11 $y = -9.95 + .23x$, $\beta_1$ is significantly different from $0 (p < 0.01)$.

5.12 $\rho_1 = 0.58$, $\hat{\rho}_2 = 0.78$ both significant at 0.05 level.

5.13 $p = 0.10$

5.14 0.6

5.15 $p = 0.322$

**Exercise 8**

8.1 (i) $1/N$ (ii) $n/N$

8.2 $(n-1)/(n+1)$, Store the largest 2 so far in the memory and replace the smaller one if $x(n+1)$ is larger. Expected # of comparisons $\doteq N + 2\ln(N)$

8.3 $2^{n-1}/n! \simeq (2e/n)^n$

8.5 No, the data may be partially ordered. The first one may be very far away from the median.

8.6 (i) 0.0027 (ii) 0.98

8.7 11.26